上 班 族 每 天 養 生 法

飲出好體質

U0100074

Drinks & Soups
FOR GOOD HEALTH
Daily Tonics for Busy Working Classes

作者簡介

　　芳姐（張佩芳）於香港大學專業進修學院修畢「中醫全科文憑」及修讀中醫全科學士多個課程。自1996年起在各大報章及雜誌撰寫食療專欄，作品散見於新晚報、大公報、壹蘋果健康網、加拿大明報周刊「樂在明廚」、及北美洲「品」美食時尚雜誌。

　　芳姐曾任「中華廚藝學院」健康美食班客座講師，現任僱員再培訓局中心、職訓局中心的陪月班導師，及「靚湯工房」榮譽食療顧問。

　　為推廣食療心得，芳姐自2007年開始在網上撰寫食療雜誌「芳姐保健湯餸」，深獲網民支持和愛戴。近年開始將其心得著作成書，著有《每天一杯養生茶》、《坐月天書》、《每天一杯養生茶2》、《懷孕天書》、《100保健湯水・茶飲》、《滋補甜品店》、《女性調補天書》、《強身・治病 孩子飲食全書》。著作迅即再版，成績斐然。

序

上班族打工仔女，無論從事體力勞動或腦力勞動，大多數精神壓力都很大，從一項調查中發現，香港人平均每周工作超過 50 小時，工時冠絕全球。工作壓力大加上食無定時，是造成亞健康體質的主要原因。亞健康體質，指的既不是完全健康，又不是處於生病的狀態，卻常會出現一些症狀：如周身酸痛、睡眠不佳、視力衰退、排便不暢、經期不順、四肢無力、腦子遲鈍、注意力無法集中等，西醫各種檢查都查不出病因，但功能已出了狀況，大多數打工仔女，正屬於這個族群。

打工仔健康欠佳的主要原因是飲食習慣沒有甚麼節制，尤其是午餐時段，進食但求速度，吃得太快、太多、太雜、太不營養，而一般食店製作的速食飯餸以煎炸食品為多，品種少，營養不全面，這種不重視營養、高熱量、高油、高鹽的餐飲文化，長期食用，自然有損健康。除了飲食習慣差，很多打工仔排便習慣亦不見得好，大多數白領不愛運動、不流汗、少飲水、淋巴循環差，體內毒素堆積，面上長期長暗瘡；而藍領一族亦會因體力虛耗大、筋骨勞損，或長期捱更抵夜，飲大量凍飲，令脾胃功能差，肝臟排毒能力減弱，因而膚色暗啞無光。

本書專為藍領、白領打工仔推薦適合的茶飲及湯品，為健康把好關，將處於亞健康的體質一點一滴慢慢改善過來。透過簡單茶飲、營養湯水，配以一些飲食常識、健康提示等，給打工仔增加更多的選擇。只要您翻開書本，定能找到合適的食譜，在享用茶飲、靚湯之餘，製作過程也可以成為生活中的輕鬆樂事。

目錄

為你的健康把好關

　　吃進肚內的東西宜格外留神。本書的湯品、茶飲是為打工仔的健康把好關,將他們處於亞健康的體質一點一滴慢慢改善過來。所以對食材要嚴加挑選,泡茶、煲湯也要用對方法,才能事半功倍。

嚴選食材

購買沒硫磺燻製的食材

　　選購食材,必須選擇優質沒硫磺燻過的,尤其是一些藥材和花草茶材料。因為食用硫磺滲入的食物後,會對人的神經系統造成損害,輕度的會出現頭昏、眼花、精神分散、全身乏力等症狀。若長期食用,嚴重的可能影響肝腎功能。

　　不良商人為了讓藥材、花草茶料防腐、防蟲蛀、保鮮、顏色好看,會將硫磺燻進去,令材料漂白、增艷,防蟲蛀,但硫磺本身性熱有毒素,對身體有害,幸而硫磺較易溶於水,所以煲湯或泡茶前將材料先用清水浸泡、換水,雖然此法不可能完全浸洗去食材中所有二氧化硫,但也可減少對人體的損害。

怎樣分辨硫磺食材

　　主要掌握三個方面:看、聞、嘗。

看:例如菊花、雪耳、百合、霸王花等,正常情況下應該是淡黃色的,如果雪雪白,看上去很漂亮,顏色太鮮艷,都可能是硫磺燻製過。

聞:各種材料有它獨特的香氣,例如雪耳有蛋白質清香味道;菊花有淡淡的菊花清香,但如果聞到有股刺鼻的酸味,則肯定被硫磺燻過的。

嘗:煲好的湯,或沖泡好的花茶,先不要加調味,可細味出湯和茶的原味,如果帶有點酸味的,就要當心了!

　　除了看、聞、嘗,一般藥材或花草茶料,存放久了必會變點顏色,甚或生蟲,例如杞子、菊花、黨參等,顏色亦會暗啞變深,如果久不變色,完全不生蟲,就要考慮可能被硫磺燻過了。

別購買過期或曾經發霉的食材

　　發過霉菌的食材含毒素頗多,屬致癌物質,亦可能引致肝硬化,絕對不能食用。有些

不良商人把已發了霉的食材，清洗、曬乾後急於低價促銷，因為霉菌已侵入食材中，就算洗過、曬過，食材很快又會長出霉菌，故會將發過霉的食材低價銷售出去。故購買時需要特別留意，蟲蛀的通常已過了期；散發霉味的多數曾發過霉，總之「便宜莫貪」，低於市價太多的東西，最好不要隨便購買。

選購藥材、花草茶，盡量去正規有信譽的店舖，最好選擇散賣的，可以拿來看看及聞聞，凡有刺鼻酸味、霉味，或有蟲蛀，顏色過分鮮艷的都不要買。倘若買包裝好的，拆開後才發覺材料有異味變了質，最好的辦法是棄之不用，千萬不要因為怕金錢損失而勉強進食，身體健康永遠要排在首要位置！

泡茶須知

沖泡花草茶其實並不複雜，與沖泡紅茶、中國茶有不少共通的原則。基本上只要把握以下幾項重點，自己在家也能完成一壺既可口又健身的養生茶。

用水

水質與水溫都是誘發花草茶茶色味道的重要因素。泡茶用水可選擇質純的礦泉水或蒸餾水，然後將水煮沸，沖泡時的水溫在攝氏95度左右，如此泡出的茶澄亮色正，味道也佳。至於用水量的多少，若以最常用的玻璃壺、瓷壺為茶具，一壺為500毫升；若在辦公室沖泡，一杯大約是250毫升。

泡茶方式

很多花草茶可用壺泡法也可用鍋煮法，通常以一種花為主，花、葉等原料因為較容易釋放出內含成分，採用壺泡法即可。若原料是果實、樹皮、根、莖等堅韌部分，採用鍋煮法較能取得其中的精華。

茶具

玻璃製品固然便於觀察顏色，但是不易保持花香，也不利於保溫，用陶具、瓷器來泡製，較用玻璃器皿更佳。鐵製或鋁製品、不銹

鋼保温杯可能會引起化學變化，尤其是材料本身有酸味的如山楂、洛神花、紫背天葵等不適合使用。

花茶配料

調製花草茶時，可加添蜂蜜、檸檬等，但不宜將所添加的蜂蜜、糖、檸檬等加得太多，以免蓋過了花草茶自然的色香味。有些植物本身味道很重，熱水一沖，味道便出來了，適宜靜置2-3分鐘飲用，否則浸泡久了，茶湯可能會變苦。

靜置時間

無論你採用何種沖泡法，都需要在沖泡後將整壺花草茶靜置2-3分鐘，讓它釋放出成分和茶香。

煲湯要訣

如沖泡花草茶般，煲湯一點也不複雜，只要把握以下的要點，煲一鍋可口香濃的湯品，絕非難事。

材料必須要新鮮

湯的味道好不好，主要看材料是否新鮮，尤其是蔬菜瓜果，煲出來的湯才可口不失營養。鮮活買回來的魚、蝦等水產，最好在宰殺後幾小時內烹煮，這樣煮出來的湯才會特別鮮甜美

味。總之，選料要鮮，用料要廣，營養才能全面。

選料搭配要適合

食物之間要講求搭配得宜，使營養素起到互補作用；同時用料要廣，營養才全面。例如用蔬果配肉類同煮，酸性的肉類與鹼性的蔬果，起「組合效應」，令湯不太肥膩及酸、鹼度適中；食物在烹煮過程中也會產生一些化學變化，使某些營養遭到破壞。例如青蘿蔔含大量維生素C，而紅蘿蔔含有維生素C分解酶，同用便會降低其營養價值；又例如用了人參、黨參等補氣中藥材，就不能用蘿蔔這類破氣食材了，因為會影響療效。

炊具選擇要適宜

煲老火湯宜用物料穩定、不易起變化的瓦煲、瓷煲、紫砂煲或玻璃煲；燉湯用紫砂煲或瓷煲較適合；滾湯則可用金屬煲，傳熱較快。湯料中有酸味的食材，容易將金屬中的物質溶解釋放出來，引致慢性中毒，故最好用玻璃製康寧煲或瓦煲等物料較穩定的煲來煲煮。

藥材熬湯要注意

使用藥材煲湯，質地清輕，氣味易散發的，需要短時間煮好，或需要用大火迫出藥效，如薄荷葉，紫蘇葉等，甚至用開水沖泡亦能出味；而質地堅硬的藥材，如黨參、人參、北芪、芡實等，就需慢火煎熬。

配水分量要適中

水的用量，水溫的變化，直接影響到湯的鮮味。用水量一般是湯料的三倍左右，蔬菜宜滾水落，可減少苦澀味；肉類、藥材宜冷水落，可讓肉中蛋白質鮮味及藥效慢慢釋出。要一次性加足水才下材料，中途加水會影響湯的濃度。

火候調校要合適

煲湯要先用武火煮滾，滾起後轉用文火慢煮，這樣才能使食品中的鮮味物質盡可能地溶解出來，讓湯既清澈，又香濃。

調味品最後才加

注意煲湯時不宜先放鹽，因鹽具有滲透作用，會使原料中水分排出，蛋白質凝固，鮮味不足。要煲好湯後才落鹽、薑、葱等調味品，以保持味道清新和鹹淡度適中。

老火湯的熬煮時間

煲湯的時間3小時以上稱為老火湯，但煲湯的時間越長，湯水中嘌呤的成分越高。對一些高尿酸、痛風及慢性腎臟患者，飲用這類老火湯，會令症狀加劇。因此，用質地較硬的藥材煲湯，可先行切細片，用清水浸半小時至1小時，令煲煮的時間縮短，可以在1.5至2小時內煮好，對健康更為有益。

老火湯的存放問題

很多打工仔沒有太多時間煲湯，故有時會煲好一大煲湯，飲剩的會隔夜存放，但部分蔬菜類、菇菌類本身含較多硝酸鹽，煮後存置時間過久會變成有害健康的亞硝酸鹽，有致癌危機，故煲這類湯最好一頓飲完，假如真的要存放第二天才飲，湯渣必須撈走，只留湯汁存放雪櫃，第二天加熱飲才會較安全。

認 識 體 質 ， 更 了 解 自 己 的 需 要

健康人士外觀上有標準可言：第一是精神好，面色好；第二是給人頭腦靈活、反應快的感覺；第三是抵抗力強，少傷風、感冒；第四是對不良環境的適應力強，耐受寒熱。

在生理方面的表現為：胃口好，睡眠好，性能力表現佳。

心理質素方面表現為：承受能力較強，情緒穩定，不易激動。

根據一項國際調查統計中指出：在所有人群中，完全健康的人只佔 15%，而患病的也佔了 15%，換句話說，介乎健康與患病中間的「亞健康」人士佔整體的70%，這情況在大部分打工仔身上所佔比率尤其明顯。

主要原因是打工仔生活壓力大，精神緊張，日常超負荷工作後得不到適當休息，睡眠不足而得不到及時緩解，體力透支未能及時補充能量……可見「亞健康」體質都是在不良環境中「熬」出來的。因此，我們更需要認識自己的體質而加以改善。而值得慶幸的是，「亞健康」體質可以透過日常的養生保健而逆轉，令人回復到健康狀態。故此，我們有需要更了解自己的體質及需要。

中醫根據基本理論，結合臨床體質調查，提出了較為常見的正常質、陽虛質、陰虛質、濕熱質、氣虛質、痰濕質、瘀血質七種臨床體質分型。

「正常質」即我們所講的健康體質。而其他六種體質，便屬「亞健康」體質了。我們可簡單的從個人形體、面色、神氣、體味、眼睛等來觀察一個人屬何種體質居多。

形態

陽虛外寒，陰虛內熱；胖人多痰濕，瘦人多內熱。

神氣

煩則多熱，靜多偏虛；神態遲鈍多痰濕，嘆氣呃逆多氣鬱。

面色

黯多瘀血，蒼白多虛；面黃無光澤多血虛，黃而油膩多濕熱。

眼睛

無神多虛，渾濁多濕；眼睛紅赤多濕熱。

舌頭

質紅多熱，苔厚多濕；舌頭胖大多陽虛，舌體偏瘦無苔多陰虛；舌色發黯有瘀斑多有瘀血積聚。

體味

體味較大，非熱即濕；經常腳臭多痰濕。

二便

尿黃多熱，便爛脾虛；尿多、夜尿頻陽虛，大便黏膩奇臭多有濕熱。

陽虛體質

特質

畏寒肢冷，虛胖肌肉鬆軟，精神不振，唇色淡白，毛髮易落，出汗多，小便清長，大便溏薄

陽虛者即使再熱的暑天，總是手腳發涼，也不敢吃太寒涼的東西。性格多沉靜及內向。易患寒症、泄瀉、水腫等症。女性明顯多於男性。長期偏嗜寒涼食物也會形成這種體質。

調養方面

可以多吃些温補腎陽的食物如羊肉、牛肉、雞肉、鴿肉、海參、鱔魚、蝦、栗子、韭菜、核桃肉、南瓜、荔枝、龍眼、櫻桃等。調味香辛料如薑、葱、蒜、花椒、胡椒等亦可多用。

太寒涼的蔬果如苦瓜、蕹菜、雪梨、馬蹄等不宜多吃。即使在炎熱的夏天，西瓜、綠豆湯、各種冰凍冷飲等亦少用為宜。

認 識 主 料

紅糖、黑糖

紅糖、黑糖皆為蔗糖製作過程中的產品，紅糖溫補，能健脾養血、祛風散寒；而黑糖保留了蔗糖較多維生素和礦物質，有活血散寒、補中益氣等功效。

薑

薑越老，薑辣素含量越多，薑辣素對消化及水液循環的停滯具有調節的作用，並可促進體內的新陳代謝，改善畏寒肢冷等不良症狀。

紅糖、黑糖

薑

黑 糖 薑 母 紅 棗 茶

暖身祛寒、促進循環。對陽虛畏
寒肢冷者特別適合,可促進血液
循環,補充體力。

飲食宜忌

此茶對女性月經暢順也有幫
助。春、秋服能預防感冒;夏季
服可止嘔止瀉;冬季服可暖身祛
寒。但要留意,久服容易積熱,損
陰傷目;陰虛內熱及痔瘡患者忌
服。

材料(2人量)

- 老薑切片、紅棗切片共
  30克

調味

- 黑糖2茶匙

做法

1 將材料放入壺內,先用
  開水沖洗一遍。
2 壺中加入黑糖,再注入
  500毫升開水,焗10分
  鐘左右可飲。

鮮淮山圓肉羊肉湯

材料（3～4 人量）

- 鮮淮山 150 克
- 新鮮羊肉 500 克
- 生薑 30 克
- 圓肉 12 克

調味

- 胡椒粉 1/4 茶匙
- 海鹽半茶匙

做法

1 鮮羊肉斬件，出水後沖洗淨，瀝乾。
2 淮山去皮後洗淨，切塊；圓肉浸洗；生薑切片。
3 將全部材料用 1.7 公升水煮至大滾，改用文火煮 2 小時，調味即可連湯料同食。

認 識 主 料

羊肉： 外國進口的急凍羊肉羶味較重，以新鮮的黑草羊最鮮味及甚少羶味。用來燜煮，羊頸肉及羊腩肉較腍滑；煲湯則可揀羊腿肉，脂肪較少。圓肉、馬蹄都適合用於羊肉湯中，因為二物都有助去除羊肉的羶味。

食 療 功 效

益氣補虛、溫補脾腎。適合陽虛四肢不溫、精神不振、面色蒼白或萎黃、頭暈心悸、大便稀溏者。

飲 食 宜 忌

本品加入了圓肉，可加強養血安神作用，氣虛血弱者亦適合飲用；如果本身容易上火，圓肉可改用馬蹄 2~3 粒代替。外感未清及陰虛內熱者不宜。

陰 虛 體 質

特質

津液不足，手足心熱，雙目乾澀，皮膚乾燥，易生皺紋。眩暈耳鳴，睡眠欠佳，小便短少，大便秘結

陰虛體質人士多體形瘦長，身體燥熱，面頰潮紅，容易失眠。這種人多數性格活潑，外向好動而性情急躁。易患神經衰弱、失眠等症。很多年輕人喜食辛辣、煎炸、燒烤食物，或嗜好煙酒，以及生活壓力大人士最易形成這種體質。

調養方面

適合甘寒涼潤、養陰生津之品。尤其是水產貝類食物如蠔豉、蛤蜊、鮑魚、瑤柱、響螺、魚肉等；蓮藕、百合、雪耳、馬蹄、香菇、各類瓜菜都適合。黑芝麻、小米、糯米、綠豆、牛奶等可常吃。

辛辣刺激的調味料如胡椒、花椒、辣椒、茴香、桂皮、老薑等易助陽生燥，盡量少用；高熱量的食物如炸洋葱圈、炸薯條、爆花米等應少吃；煙酒最好戒掉。

低溫脫水羅漢果

認 識 主 料

羅漢果

羅漢果較蔗糖甜 300 倍，卻
具有降血糖的作用，可作為
糖尿病、高血壓、高血脂
和肥胖症患者之首選的天
然甜味劑。選購時以顆粒
大、皮色均勻、搖之不響者
為佳。現時市場有低溫脫水
羅漢果出售，泡出來的茶色
較清，味道甘甜清香。本頁
介紹的茶水是用低溫脫水
羅漢果泡的。

羅 漢 果 麥 冬 茶

食療功效

滋陰生津、清熱利咽。對陰虛咽乾喉痛、燥咳、肺熱咳嗽、腸燥便秘者有幫助。

飲食宜忌

本茶對常用嗓子工作者如教師、歌唱家、經紀推銷員等最合適，可防慢性咽炎；但體質虛寒者不宜。

材料（2人量）

· 羅漢果6克
· 麥冬10克

做法

1 將羅漢果、麥冬放入壺內，用開水先沖洗一遍。
2 再注入500毫升開水，焗15分鐘可飲。

沙 參 玉 竹 水 鴨 湯

材料（3~4 人量）

- 沙參 15 克
- 玉竹 12 克
- 雪耳 5 克
- 杞子 5 克
- 陳皮 1 塊
- 水鴨 1 隻

調味

- 海鹽半茶匙

做法

1 水鴨劏洗淨，斬件後出水。
2 沙參、玉竹、陳皮、杞子、雪耳浸洗，雪耳去蒂。
3 將全部材料（除杞子外）放入煲內，用 1.5 公升水先煮至大滾，改用文火煮個半小時，加入杞子煮 5 分鐘，調味即可連湯料同食。

認 識 主 料

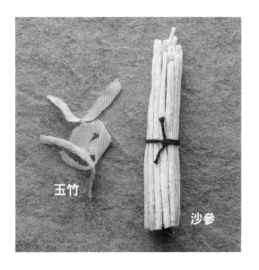

沙參：有分南沙參和北沙參，以幼身結實的北沙參功效較佳。

玉竹：買時宜拿上手聞聞，如色太白、味帶酸者可能有硫磺燻過，不要買。

水鴨：滋陰補虛，煲湯不太肥膩，如買不到急凍水鴨，可用老鴨半隻代替。

食療功效

清熱養陰、潤燥生津。適合陰虛火旺、五心煩熱、皮膚乾燥、肺燥咳嗽者飲用。

飲食宜忌

本湯對經常捱夜，虛不受補者亦很適合。但風寒咳嗽及脾虛便溏者慎服；痰熱咳嗽者忌服。

痰濕體質

特質

腹肌鬆軟，形體肥胖，面部多油，多汗且黏，胸悶，痰多，容易困倦，舌體胖大，小便不多，大便不實

痰濕人士多見形體肥胖，皮膚油脂較多，多汗且黏，胸悶、痰多，但性格偏於溫和，忍耐力強。由於痰濕內阻，易患高血壓、糖尿病、冠心病、中風等症。少做運動、愛睡覺、喜甜食，生活安逸的中、老年男性，易形成這種體質。

調養方面

適合用健脾益腎、清利化濕的食物如冬瓜、薏米、扁豆、紅豆、山楂、花生、海帶、海蜇、洋葱、蘿蔔、芹菜、各種瓜果、蓮藕、椰菜等。忌油膩食物，甜品、酒類不宜常用，不能暴飲暴食，應限制鹽、糖及油的用量，飲食盡可能清淡。

認識主料

雲苓

又稱茯苓。黑褐色表皮稱茯苓皮，長於利水消腫；內部淡紅色者稱赤茯苓，長於清利濕熱；再內白色者稱白茯苓（或雲苓），長於健脾滲濕；中間心白色的為茯神，用於安神。茯神中間有粒「芯」的，是抱住松根而生的野生茯神，安神功效更佳。

扁豆衣

健脾和胃，消暑化濕。以囊殼完整、色黃白、不帶種仁者為佳。

白朮

健脾益氣，燥濕利水，止汗，安胎。以切片淺黃色、乾燥無蟲蛀者為佳。

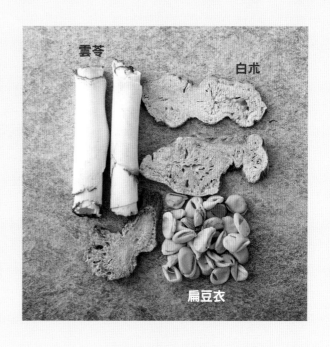

雲苓白朮
扁豆衣水

材料（2 人量）

· 雲苓、白朮、扁豆
  衣、甘草共 30 克

做法

將材料浸洗，用 700
毫升水煮 30 分鐘即可
供飲。

食療功效

健脾益氣、利水滲濕。適
合濕重痰多、身體肌肉鬆
軟、睡覺流口水、四肢水
腫者飲用。

飲食宜忌

本湯對感冒後脾胃功能未恢
復，又想進補者很適合，先健
脾祛濕，進食補益品會較易吸
收；陰虛無濕熱者不宜服。

冬瓜赤小豆薏米排骨湯

冬瓜

薏米

材料（3 ~ 4 人量）

- 冬瓜 500 克
- 赤小豆 50 克
- 薏米 50 克
- 蜜棗 2 粒
- 排骨 300 克

調 味

- 海鹽半茶匙

做 法

1 排骨斬件，洗淨出水。
2 赤小豆、薏米浸洗；冬瓜連皮洗淨，切塊。
3 將全部材料用 1.7 公升水先煮至大滾，改用文火煮 2 小時，調味即成。

認 識 主 料

冬瓜：能清熱化痰，利水消腫，然而冬瓜皮不但能清熱利水消腫，更能治療糖尿病，故煲冬瓜湯宜連皮一齊煲。

薏米：有生、熟薏米之分，生薏米利水滲濕功效較佳；熟薏米寒性減少，功能健脾止瀉。

消暑祛痰，清熱祛濕。適合
身體濕熱不適、皮膚濕疹、口
乾口渴、痰多、小便不暢
順、尿液黃赤者飲用。

本湯對痛風及筋脈
痙攣者都有幫助；但
脾胃虛寒、夜尿多
者不宜服。

氣虛體質

特質

肌肉鬆軟，雙目無神，氣短懶言，易出汗，頭暈，健忘，不耐寒熱，易患感冒，小便偏多，大便不成形或便後仍覺未盡

氣虛體質人士活動能力不足，說話沒勁，氣虛陽弱，性格內向，情緒不太穩定，膽子細小。容易呼吸短促，經常神疲乏力，易出虛汗，不耐勞動，易患傷風、感冒、胃下垂、脫肛等症。失業人士、學生和長期從事體力勞動者也易氣虛。

調養方面

宜多吃具益氣健脾作用的食物如：雞肉、泥鰍、鱔魚、鵪鶉、牛肉、扁豆、豆腐、糯米、小米、香菇、葡萄乾、蜂蜜等。忌耗氣、破氣之品如蘿蔔、馬蹄、山楂、蕎麥、金橘、綠茶等。飲食宜清淡，太多辛辣厚味的香辛料，過食易動火耗氣。

認 識 主 料

黃蓍

能補氣升陽，是很好的補氣藥，以表皮皺紋稀少，質堅而綿、粉性足、味甜者為佳。

杞子

以寧夏出產者質較優，其特點是皮薄、肉多、核少、味甜。購買時要留意杞子頂部有個淺黃色的果柄圈，如全粒皆鮮紅色，多數是染色杞子，不要購買。

杞子

黃蓍

黃 蓍 杞 子 紅 棗 茶

材料（2 人量）

- 黃蓍 15 克
- 杞子 6 克
- 紅棗 6 粒

做法

1　黃蓍切碎，與杞子用水浸洗；紅棗去核，切片。
2　將材料放入壺內，用開水沖洗一次，再注入 500 毫升開水，焗 15 分鐘可飲用。

食療功效

益氣固表、強壯體質。適合氣虛乏力、雙目無神、容易傷風、感冒、汗多者飲用。

飲食宜忌

本茶在感冒多發季節有強肺及抗疫功效，全家老少可飲。但邪氣盛、有熱毒者不宜。

黨參杞子燉花膠

食療功效

補中益氣、固腎益精。適合氣血兩虛、肺腎虛損、氣短心悸、神疲乏力、面色蒼白、頭昏眼花、腰膝無力者飲用。

飲食宜忌

本品男女可服，更適合手術後人士服用。常服能令肌膚細嫩，面色紅潤。但脾胃氣滯、肝火盛者及外感未清者不宜。

認識主料

黨參：以山西上黨地區生產的黨參為上品。有增強免疫能力、擴張血管、改善微循環、增強造血等功能。選購以質稍硬略帶韌性，皮紋清晰及淡棕色，肉部淡黃色。有特殊香氣，味微甜者為佳。

材料（2~3人量）

- 黨參 15 克
- 杞子 6 克
- 生薑 3 片
- 紅棗 6 粒
- 瘦肉 250 克
- 浸發花膠 120 克

調味

- 海鹽半茶匙

做法

1　瘦肉洗淨切片，與浸發花膠同出水；紅棗去核；黨參、杞子浸洗。

2　將全部材料放入燉盅內，注入 600 毫升開水，隔水燉 2 小時，調味即成。

花膠

黨參

花膠：滋陰養顏，補血，補腎，強壯機能。以完整、無白斑、無破損、煲起不腥不溚不溶化者為佳。
附圖的是花膠公，呈人字紋。

血瘀體質

特質

體形偏瘦，面色晦暗，易出色斑，易患痛症，眼眶黯黑，皮膚乾燥，頭髮易落，痛經閉經，唇色暗淡，舌下靜脈曲張

血瘀體質人士多形體消瘦，性情急躁，頭髮易脫落，肌膚乾燥粗糙，色素沉着，又易患各種痛症、出血症、中風、冠心病等。女性多見痛經、閉經或崩漏。腦力工作者、過於安逸人士及喜冰凍冷食的女性易出現血瘀。

調養方面

宜多食蓮藕、黑豆、陳皮、海藻類、蘿蔔、猴頭菇、蘑菇、香菇等理氣食物，可適度飲些葡萄酒、黃酒、白酒；平時宜飲些玫瑰、茉莉花茶等具疏肝解鬱的茶飲，不宜過食寒涼的食物。具收斂固澀的食物如芡實、蓮子、浮小麥等亦要少用。

認 識 主 料

山楂

具有降血脂、降血壓、強心、抗
心律不齊等作用，同時也是健脾
開胃、消食化滯、活血化瘀的
良藥。山楂有分南山楂和北山
楂，南山楂為野生山楂，質地堅
硬，核大，果肉薄色白，皮棕紅
色。氣微，味酸澀。北山楂果肉
深黃色至淺棕色，氣微清香，味
酸微甜。療效以北山楂較佳，以
個大、皮紅、肉厚者為上品。

山 楂 蜜 茶

材料（2人量）

· 山楂 30 克

調味

· 蜂蜜 1 湯匙

做法

1 將山楂放入壺內，用開
   水沖洗一遍，將水倒走。
2 再注入 500 毫升開
   水，焗 10 分鐘左右，加
   入蜂蜜即可飲用。

食療功效

消食化積、活血散瘀。適合面
色晦暗，長有色斑、血脂高、血
壓高、食滯人士飲用。

飲食宜忌

本品能去油膩肉積，但山楂味
太酸，易傷及於胃，而蜂蜜益
陰潤燥和補脾氣，能舒緩對
胃的刺激，長者可以加些紅茶
或普洱茶同泡，更能促進消
化。但孕婦、血糖過低、脾胃
虛弱及胃潰瘍者忌服。

益母草黑豆雞胸肉湯

材料（2～3人量）

- 益母草 30 克
- 青仁黑豆 50 克
- 紅棗 6 粒
- 生薑 2 片
- 雞胸肉 150 克

調味

- 海鹽半茶匙

做法

1. 雞胸肉切片，出水；益母草洗淨。
2. 黑豆浸軟；紅棗去核。
3. 將全部材料放入瓦煲內，用 800 毫升水煮至大滾，改用文火煮 1 小時，調味即成。

益母草：新鮮青嫩的益母草為童子益母草，是第一年生長的，可以當蔬菜食用，可調經、降血壓、利尿消水腫，並有養血功效。但療效以二年生的乾品益母草較佳。

食 療 功 效

活血祛瘀、行氣補虛。適合面色晦暗、唇色暗淡、月經不調、水腫尿少者飲用。

飲 食 宜 忌

本品能去瘀生新，並能抗氧化、防衰老，具有相當不錯的益顏美容功效。但婦女來經期間不宜服，孕婦禁服。

黑豆：青仁黑豆肉青色、皮黑色，含有的花青素能清除體內自由基，尤其是在胃酸的環境下，抗氧化效果好，常食能延緩衰老。中醫認為青色入肝、黑色入腎，故青仁黑豆肝腎雙補。

氣 鬱 體 質

特質

形體偏瘦，神情抑鬱，胸脅脹滿，乳房脹痛，心悸失眠，咽間有異物感，小便正常，大便偏乾

氣鬱體質情志內鬱不暢，性格內向不穩定，多愁善感，敏感多疑、易患抑鬱、驚恐、臟躁、神經衰弱、失眠等症。對精神刺激適應能力較差、性格內向的年輕人、體虛長者，及更年期女性較易出現氣鬱。

調養方面

宜多吃行氣、理氣、解鬱、調理脾胃功能的食物，如蘿蔔、大豆、桔橘類、柚子、麥米、蕎麥、金針、木耳、海帶、山楂、葡萄乾等。平時可以飲些牛奶及沖泡各種花草茶飲品，可少量飲紅酒、糯米酒以活動血脈、提振精神。避免肥甘厚味，以防氣機壅滯。高糖分的汽水、瓶裝的果汁及冰凍冷飲會加重氣滯狀況，盡量少飲。

合歡花

於 6 月花初開時採摘,有很多絨毛,故乾燥花序呈團塊狀,有如棉絮;花未開時採的花蕾,稱合歡米。兩者食療功效相若。有些中藥店以大朵木蘭科的夜合花作合歡花出售,這種棕色大朵的夜合花,其作用是行氣祛瘀、止咳止帶,亦可治肝鬱氣痛,但解鬱療效遠不及合歡花。

佛手柑

能舒肝理氣,和胃止痛。以外皮鮮黃色,質硬而脆,氣香無受潮者為佳。

佛手柑

合歡花

合 歡 花 佛 手 茶

材 料（2 人 量）

- 合歡花 5 克
- 佛手柑 10 克

調 味

- 蜂蜜 1 湯匙

做 法

1　合歡花、佛手柑同放入壺內，用開水沖洗一遍，將水倒走。
2　再注入 500 毫升開水，焗 10 分鐘，調入蜂蜜即成。

食 療 功 效

舒解鬱結、緩和緊張。適合胸脇滿悶、憂鬱焦慮、敏感多疑、失眠、健忘者飲用。

飲 食 宜 忌

本品常用於心神不安、憂鬱失眠，常服能舒緩緊張，寧心安神，減輕疲勞。
風火眼疾、視物不清都可用，但孕婦不宜。

金針雲耳肉片湯

材料（3～4人量）

- 金針 15 克
- 雲耳 6 克
- 雞蛋 2 個
- 瘦肉 250 克
- 芫茜 2 棵

調味

- 海鹽半茶匙

做法

1. 金針、雲耳浸洗，金針打結，雲耳去蒂；瘦肉切片，與金針、雲耳同汆水，撈出瀝乾。
2. 雞蛋打散；芫茜洗淨，切碎。
3. 燒熱 700 毫升水，將金針、雲耳、瘦肉放入煮 10 分鐘，加入蛋漿及芫茜碎，調味後再滾起即成。

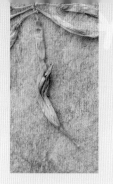

金針：含「秋水仙鹼」，這種物質在鮮品中含量最多，食用過量鮮金針會出現噁心、嘔吐，甚至休克的中毒症狀。乾金針秋水仙鹼成分大大減少，但金針在烘乾過程中，商人為保持金針的鮮黃或橘黃色，會用硫磺燻蒸，雖然在最後階段會進行脫硫，但為了食用安全，最好浸水 1 小時，汆燙 1 分鐘，可減少硫磺殘留量。選購金針時，以氣清香、乾燥無酸味者為宜，黑褐色的不要買。

食 療 功 效

疏肝解鬱，滋補安神。適合情志內鬱不暢、面色晦暗、心悸失眠、神經衰弱者服用。

飲 食 宜 忌

本品清香味美，營養豐富，老幼可服。可清除肺熱、柔和肝氣，任何體質人士可服。

濕熱體質

特質

身體易上火發炎，面色黃而油膩，易興奮緊張，眼熱分泌物多，口乾舌燥，喜喝冷飲，小便短赤，大便乾結黏滯，腳臭薰人

濕熱體質人士多形體偏胖，面泛油光，易生痤瘡粉刺，口氣不佳，身重困倦，胸脅脹悶，眼睛紅赤，皮膚紅腫熱癢。男易陰囊潮濕搔癢，女易帶下色黃。很多久居濕地，喜食肥甘，或長期飲酒，或性格急躁易怒者易形成這種體質。

調養方面

宜多吃涼性降火、解暑化濕的食物，如苦瓜、冬瓜、黃瓜、綠豆、芹菜、蓮藕、紫菜、海帶、豆腐、西洋菜、西瓜、雪梨、馬蹄、蘆薈、鴨肉、鯽魚、蛤蜊等。綠茶、花草茶也對改善體質很有幫助。

飲食忌肥甘油膩、易上火食物，如牛肉、羊肉、酒類、韭菜等；辛辣食物如辣椒、芥辣、胡椒、花椒、生薑、肉桂等，會助長濕熱，少食為宜。

認 識 主 料

荷葉

有良好的消暑解熱、降血脂、降膽固醇和減肥作用，鮮荷葉氣清香，用來煲粥服減肥效果甚佳。坊間有些炒製過，像茶葉一樣的荷葉茶，寒涼之性稍減，體質偏寒者都可以飲用。

苦瓜乾

是野生小苦瓜曬乾而成。苦瓜乾不僅可改善高脂飲食誘發的糖尿病，而且能減肥、降血脂，特別是對預防腹部脂肪堆積格外有作用，是減肥瘦身佳品。

荷葉

苦瓜乾

荷 葉 苦 瓜 茶

材 料（2 人 量）

· 乾荷葉 10 克
· 苦瓜乾 20 克

做 法

1 將乾荷葉、苦瓜乾放
   入壺內，用開水沖洗
   一遍，將水倒走。

2 再注入 500 毫升開
   水，焗 10 分鐘即成。

食療功效

清泄胃火、消脂降壓。適合濕熱、肥胖、消渴、目赤腫痛、生熱痱、胃火盛生痱滋、小便短赤及痰濕體質者飲用。

飲食宜忌

本茶味微苦，最適合糖尿病者、濕熱、肥胖水腫，在高溫環境作業人群飲用；但體質虛寒及體瘦氣血虛弱者慎服。

紫菜豆腐蛋花湯

材料（3～4人量）

- 紫菜 1 小撮
- 豆腐 2 磚
- 雞蛋 2 個
- 薑絲 1 湯匙
- 葱花 1 湯匙

調味

- 上湯 800 毫升
- 海鹽半茶匙

食 療 功 效

益氣健胃、消脂降壓。適
合濕熱、肥胖，氣虛體
質、口臭口渴、腸胃積熱
及熱病後調養者食用。

飲 食 宜 忌

本品營養豐富，是濕熱及
肥胖者補充營養的食療
佳品。但痛風患者及脾
虛、脾胃虛寒者不宜。

認 識 主 料

豆腐：買回來後如不是即時烹煮，可用
淡鹽水浸，能保持豆腐香氣，防止變質。

紫菜：具有化痰軟堅、清熱利水、補腎
養心的功效。紫菜含碘量高，可治療因
缺碘引起的甲狀腺腫大，並可降低血清
膽固醇。以色澤紫紅、無泥沙雜質、乾
燥者為佳。

豆腐

紫菜

做 法

1　豆腐洗淨，切粗絲；紫菜沖洗；雞
　　蛋打散。
2　燒熱上湯，將豆腐、薑絲煮滾，加
　　入紫菜、蛋汁，最後加海鹽及葱花
　　即成。

預防肥胖的食療

肥胖原因複雜，有些是遺傳性的，父母中有一人肥胖，子女有 40% 可能出現肥胖；倘若父母皆肥胖，子女肥胖的機率高達 70~80%。社會環境因素亦是造成肥胖主因，各種美食當前，很易讓精神緊張、情緒上不穩定的上班族以吃來作宣泄，加上大部分打工仔缺少運動，更易導致肥胖。

很多打工仔因為白天工作忙，午餐隨便吃點東西，晚上又可能要加班，下了班心情放鬆了，會感到特別肚餓，於是吃得比較多，但吃後已接近睡覺時間，沒有讓食物完全消化，食物不消化就等於攝取了更加多熱量，熱量只能轉變成為脂肪，故太夜進食的打工仔女，都較容易發胖。

男性脂肪分布以頸部及軀幹、腹部為主，四肢較少；女性則以胸部、腹部、臀部及四肢為主。無論男女，肥胖的主因是吃了太多「油」，而這種油脂含飽和脂肪，多來自牛油、豬油及各種肉類等等，但魚肉反而含豐富的不飽和脂肪。蛋白質也是一種能產生熱量的物質，如果吃多了，攝入的熱量超過了人體需要量，儲存起來會轉化成脂肪。含蛋白質的食物多為肉類、蛋類、奶類、黃豆類、粗糧，蔬菜和水果也有少量的蛋白質。

不少人以為澱粉質最易使人發胖，故盡量不沾米飯。有些女性更會以純吃生果蔬菜或素食來減肥，這種偏食的飲食方式，營養欠缺均衡，長期下來定會把身體正常機能搞亂，對健康有損。其實健康減肥飲食，可奉行目前醫學界提倡的金字塔飲食方程式，以澱粉質為基層最重要食物，順序而上是蔬果、肉類，而油類則放於金字塔最頂層。若堅持健康飲食法，再配合適當運動，持之以恆，定能見到效益。

認 識 主 料

玳玳花

春夏開白花，香氣濃郁；果實扁球形，冬季為橙紅色，翌年夏季又變青，故稱「回青橙」。因果實有數代同生一樹習性，亦稱「公孫桔」。選購以花瓣米黃色、乾身及氣香者為佳。

柑桔蜜

在大型超市有售。平時單獨沖泡，可舒緩虛火喉痛、胃脹不舒，適合咳嗽痰多而體質偏寒者；柑桔蜜同時可以美顏潤膚、潤腸通便、養心安神。

柑桔蜜　　　　　玳玳花

玳 玳 花 柑 桔 蜜 茶

材料（2人量）
- 玫瑰花6克

調味
- 柑桔蜜1湯匙

做法
1. 玫瑰花放入壺內，用開水沖洗一遍，將水倒走。
2. 再注入500毫升開水，焗5分鐘，調入柑桔蜜即可服用。

疏肝理氣、健胃消脂。適合氣鬱、胸中痞悶、脘腹脹痛、嘔吐少食及脾胃失調而肥胖者飲用。

飲食宜忌

本品清香，常服能促進血液循環，養顏美膚，可以說是一種美容茶。任何人士均可服，但孕婦忌服。

丁香茉莉花茶

材料（2人量）

- 丁香 3 克
- 茉莉花 5 克
- 綠茶 3 克

做法

1 將材料放入壺內，先用開水沖洗一遍，將水倒走。

2 再注入 500 毫升開水，焗 5 分鐘可飲用。

丁香：丁香為芳香健胃劑，可緩解腹部脹氣。丁香以花蕾及果實入藥。花蕾稱公丁香，果實稱母丁香。公丁香助陽之力較強。

食療功效

本茶能温補腎陽、纖體瘦身。適合陽虛體質、肥胖、精神壓力大、有口氣者飲用。

飲食宜忌

本品健胃，能緩解胃腸脹氣，更有助清新口氣及抗抑鬱。但陰虛內熱、口苦口乾及燥結便秘者不宜。

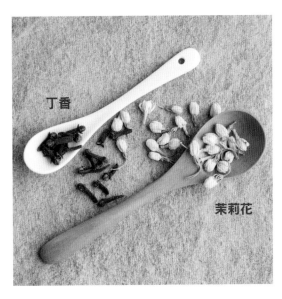

丁香

茉莉花

茉莉花：茉莉花的香氣可穩定情緒，還可清新口氣，調節內分泌，潤澤膚色。選購以花蕾緊結，較圓而短小，香氣濃者為佳。

肉桂麥芽茶

材料（2人量）
- 肉桂粉半茶匙
- 炒麥芽 30 克

調味
- 紅糖 2 茶匙

做法
1 炒麥芽沖洗淨，瀝乾。
2 用 600 毫升水煮 15 分鐘，加入肉桂粉及紅糖，糖溶即可飲用。

食療功效
溫中健胃、活血消脂。適合食積不消、脘腹脹痛、脾胃虛寒、腰膝冷痛及陽虛體質者飲用。

飲食宜忌
本品芳香甜美，糖尿病者只要不加糖都可以常服，有助健胃降血糖。但陰虛火旺、有實熱及孕婦、哺乳產婦忌服。

認識主料

麥芽：有分生麥芽、炒麥芽和焦麥芽。生麥芽健脾和胃、疏肝行氣、回乳；炒麥芽行氣消食、回乳；焦麥芽消食化滯。

肉桂：是桂皮磨成的粉，它是一種芳香香料，主要作用是溫腎、散寒止痛，對好食生冷品人士最適合。肉桂取自肉桂樹的樹皮，越接近樹幹中心的樹皮所製成的肉桂粉品質越高。

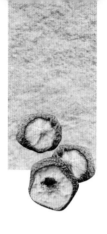

山楂決明茶

材料（2 人量）

- 山楂 15 克
- 炒決明子 20 克

調味

- 紅糖 2 茶匙

做法

1. 山楂、炒決明子放入壺內，先用開水沖洗一遍，將水倒走。
2. 再注入 500 毫升開水及紅糖，焗 15 分鐘可飲。

食療功效

消食導滯、活血消脂。適合吃太多肉食或油膩而消化不良、肥胖症、高血壓症、冠心病及血瘀體質者服用。

飲食宜忌

本品可降脂減肥，同時可以清除血中囤積的脂肪。一般人皆可服用，但便溏泄瀉、消化性潰瘍、胃酸過多者及孕婦不宜。

認 識 主 料

山楂：能開胃消積、活血散瘀、化痰行氣。有分南山楂和北山楂，南山楂皮紅肉白；北山楂皮紅肉黃，以北山楂功效較佳。山楂不宜同海鮮、人參、檸檬同食。

決明子：能清熱明目，潤腸通便。可治療高血壓、肝炎、習慣性便秘。決明子炒後可減緩滑腸作用，且質較鬆脆，易於粉碎和煎出有效成分。決明子以棕褐色有光澤、乾燥無蟲蛀者為佳。

淡菜節瓜薏米排骨湯

材料（2 人量）

- 淡菜 60 克
- 節瓜 2 個
- 生薏米 30 克
- 排骨 250 克
- 生薑 3 片

調味

- 海鹽半茶匙

做法

1　淡菜浸洗；節瓜刮去外皮，切塊；薏米浸洗；排骨出水。
2　燒熱 1.5 公升水，放入全部材料，煮 2 小時，加入調味即成。

食療功效

健脾開胃、清熱消脂。適合氣血不足、營養不良、水腫、高血壓、脾虛多汗、二便不暢者飲用。

飲食宜忌

本品鮮甜可口，健脾益腎、利大小便。任何體質可服，但孕婦不宜吃薏米，可改用眉豆 30 克代替。

認 識 主 料

淡菜：能補肝腎、益精血、消癭瘤。選購以顆粒大小均勻、色澤黃或橙黃、略有光亮、乾燥有香氣者為佳。

節瓜：有清熱、解暑、利尿、消腫等功效，是四季可食之瓜菜。節瓜以瓜皮顏色翠綠，瓜身多毛、呈光澤者為佳。

合掌瓜眉豆鯇魚尾湯

材料（3~4人量）

- 合掌瓜 2 個
- 眉豆 50 克
- 陳皮 1 塊
- 鯇魚尾 1 條

調味

- 海鹽半茶匙

做法

1. 合掌瓜去皮，切塊；眉豆、陳皮浸洗。

2. 鯇魚尾洗淨，用少許油煎香。

3. 燒熱 1.5 公升水，加入全部材料煮 1 小時，加入調味即成。

認 識 主 料

合掌瓜：在瓜類中營養較全面，有助增進食慾並維持神經系統運作；同時能強化人體免疫功能。另其纖維素含量亦不少，提供飽足感之餘，更有助於穩定血糖和血脂水平。

食 療 功 效

益脾補腎、利水消腫。適合消化不良、水腫、暑熱頭脹頭昏、噁心、煩躁者服用。

飲 食 宜 忌

本品清甜味美，老少皆宜，任何體質者皆可服用。痛風患者不宜吃眉豆，可改用薏米 30 克代替。

眉豆：對脾虛有濕、體倦乏力、便溏者有益。但眉豆中有一種凝血物質及溶血性皂素，必需煮至熟透才能吃，如煮不熟，在食後 3 至 4 小時，部分人會有頭痛、頭昏、噁心、嘔吐等中毒反應。

牛蒡雪耳粟米瘦肉湯

材 料（3～4人量）
- 鮮牛蒡 150 克
- 雪耳 10 克
- 粟米 2 支
- 生薑 2 片
- 瘦肉 250 克

調 味
- 海鹽半茶匙

做 法

1　牛蒡連皮洗擦乾淨，切段；粟米去衣，洗淨，切塊；雪耳浸軟，去蒂。

2　瘦肉切片，出水；將全部材料放入煲內，用 1.5 公升水煮 1 小時，調味，可連湯料同食。

068

認 識 主 料

牛蒡：乃淨化血液之佳品，選購以細長、形態筆直無分岔、整體粗細均勻一致者較佳；表皮最好是淡褐色且不長鬚根，質地細嫩而不粗糙。表皮粗糙且鬚根長，表示肉質鬆散，吃起來的口感也會較差。

食 療 功 效

消脂減肥、利尿消腫。適合肥胖、水腫、便秘、尿黃、尿少、三高症及痰濕質人士服用。

飲 食 宜 忌

本品清甜好味，能淨化人體廢物積存、減少毒素及防癌、防中風；但脾胃虛寒者少服。

雪耳：含豐富蛋白質和氨基酸，能養陰補肺，有「窮人燕窩」之稱。選購以乾燥、色澤淡黃、肉厚、整朵、無刺激性氣味為佳，過白或過黃，並附有刺鼻氣味的都不適宜。

認 識 主 料

蘆筍：有「蔬菜之王」的美譽。因為蘆筍味美又營養豐富，有防癌抗輻射、提高身體免疫力的功效。選購以筍尖葉片緊密飽滿；筍莖形狀要圓，表皮顏色鮮亮無皺縮；基部切口無變色；聞時沒腐臭味為佳。

蘆筍竹笙蚌肉湯

材料（3 ~ 4 人 量）

- 蘆筍 150 克
- 竹笙 20 克
- 杞子 3 克
- 新鮮連殼蚌肉 300 克
- 薑絲 1 湯匙

調 味

- 海鹽半茶匙

做 法

1 蘆筍去皮，洗淨切段；杞子浸洗；竹笙浸軟後剪去頭、尾，汆水後切段；蚌肉出水，起肉。

2 燒熱 1.2 公升水，加入全部材料，煮 20 分鐘，加入調味即可連湯料同食。

竹笙：甚為「刮油」，因含有大量粗蛋白與谷氨酸，能有效減少腹壁脂肪囤積，更有止痛、補氣、降血壓及降膽固醇作用。選購以色澤淺黃、肉厚質柔軟、菌朵完整、氣香無刺鼻酸味者為佳。

食療功效	飲食宜忌
健脾利水、滋陰清熱。適合肥胖、高血壓、心臟病、眼目昏花、耳鳴耳聾、陰虛內熱者食用。	本品鮮甜美味,能除煩解熱明目。但脾胃虛寒、外感未清、陽虛便滑者不宜。

預 防 三 高 的 食 療

三高是指高血脂、高血壓、高血糖，是現代社會發展出來的「富貴病」，可以單獨存在，也可以互相關連。例如糖尿病患者很易同時患上高血壓和高血脂症；而高血脂又易形成動脈粥樣硬化，加劇引致血壓升高。打工仔女工作忙碌，精神緊張，很多時存有 Work hard Eat hard 這種心態，放工後盡情吃喝，大杯酒大塊肉作為減壓、消除疲勞的方式，而正是因為這種不懂節制的飲食法，日漸形成了三高症而不自知。

早期患上高血壓可能只出現頭痛、頭暈、眼花、心悸、記憶力減退、注意力不集中情況，漸漸出現手腳麻木、疲乏無力，血壓持續在較高水平，後期腦、心、腎器官開始受損，這些器官受損早期可能察覺不到症狀，但最後卻導致功能障礙，甚至出現衰竭。

糖尿病是由於體內胰島素不足，使體內的碳水化合物、脂肪、蛋白質等營養代謝異常，其症狀為「三多一少」，即多尿、多飲、多食及體重下降。此病可引起多種併發症，嚴重時可引起全身性疾病，導致殘廢，致盲，甚至死亡。

高血脂又稱血脂異常，是血中膽固醇或甘油三酯過高或高密度蛋白膽固醇過低。由於血脂過多，易造成血稠，在血管壁上沉積，逐漸形成動脈粥樣硬化，堵塞血管，使血流變慢。這種情況如發生在心臟，會引起冠心病；發生在腦部會出現腦中風；如果堵塞在眼底，視力下降甚至失明；如發生在腎臟，就會引起腎動脈硬化、腎功能衰竭；發生在下肢，會出現肢體壞死、潰爛等。此外，高血脂亦可引發高血壓，誘發腦退化症等疾病。

因此，打工一族要合理調節飲食，減少大杯酒大塊肉這種減壓方式，更要戒煙，控制飲酒，同時最好經常做些慢跑、游泳一類的帶氧運動，遠離三高。預防勝於治療，預防三高症，就要多了解三高症的危害，對其預防措施多懂一些；減少吃高熱量、高脂肪的食物，勤做運動。而更重要的，是學習精神放鬆，泡些養生花草茶，煲些保健湯水，用合理的飲食來調節，三高症就不會來襲擊了。

雪 菊

以高海拔種植的較優質。高山種植的雪菊花蕾收緊，花瓣顏色金黃鮮艷，花瓣較脆，泡出來的茶色絳紅清透，味清香；低海拔山地種植的雪菊，花蕾蓬鬆，顏色暗紅少光澤，花瓣較軟，泡出來的茶色暗黃，微有酸苦味。

蕎 麥

有分甜蕎麥及苦蕎麥兩種。苦蕎麥被譽為「五穀之王」，由於經過烘製過程，色澤金黃，又稱為金蕎麥。金蕎麥富含有大量的盧丁和維生素PP，較甜蕎麥多。蕎麥能美容養顏，減少細紋；健胃排毒，幫助減輕體重。且可預防「三高」。甜蕎麥超市或雜貨店有售，苦蕎麥則要在有機店舖或花茶店購買。

蕎麥

雪菊

雪 菊 蕎 麥 茶

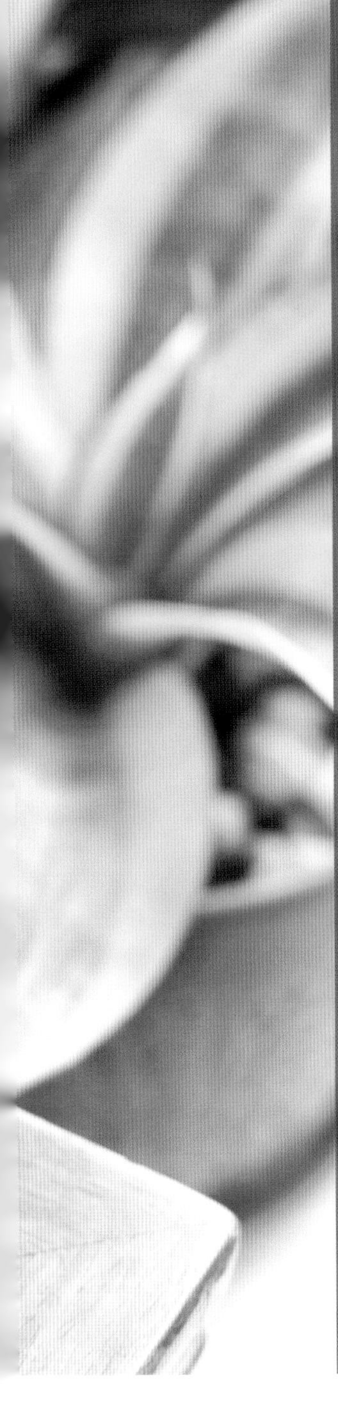

材料（2人量）

- 雪菊10克
- 炒蕎麥30克

做法

1 雪菊、炒蕎麥放入壺內，先用開水沖洗一遍，將水倒走。

2 再注入500毫升開水，焗15分鐘可飲用。

食療功效

消脂降壓、下氣消積。適合腸胃積滯、食慾不振、高血壓、高血脂、高血糖、冠心病及陰虛體質者飲用。

飲食宜忌

本品香味濃郁、茶色絳紅而清透，適合肥胖及三高症者常服。但低血壓、體質虛弱及平素過敏者慎服。

番石榴綠茶

材料（2 人量）

· 番石榴乾 30 克
· 綠茶 5 克

做法

1　番石榴乾放入壺內，先用
　　開水沖洗一遍，將水倒走。
2　注入 500 毫升開水，加入
　　綠茶，焗 7 分鐘可飲用。

認 識 主 料

番石榴乾：能軟化血管、降血脂、降血糖、降膽固醇，同時有助消除疲勞。可以在山草藥檔、花草茶店購買，以乾身、無蟲蛀、氣香者為佳。

食 療 功 效

軟化血管、防治三高。適合肥胖、三高症、口臭、牙肉腫痛及痰濕體質者飲用。

飲 食 宜 忌

本品微苦但飲後回甘，中、老年人最合飲用。但脾胃虛寒及大便秘結、瀉痢積滯未清者忌服。

綠茶：龍井、碧螺春、竹葉青、毛峰茶等都屬綠茶。綠茶能提神清心、去膩減肥、生津止渴、除煩、醒酒等。從性質來看，綠茶屬涼性，而紅茶偏溫，所以胃腸比較弱的人宜用紅茶。

洛神花蘋果茶

材料（2人量）
· 洛神花6朵
· 蘋果1個

調味
· 蜂蜜1湯匙

做法
1　蘋果洗淨去皮，切粗粒。
2　洛神花、蘋果粒用600毫升水煮15分鐘，飲時調入蜂蜜即成。

認 識 主 料

洛神花：有益於平衡血脂，增進鈣吸收，促進消化，並有助消除疲勞。購買時以花形完整，暗紅色透點鮮紅，乾身微有酸味者為佳。

食 療 功 效

美容嫩膚、消脂降壓。適合肥胖、高血脂、高膽固醇、高血糖、容易疲倦、煩熱口渴、過量飲酒及痰濕體質者服用。

飲 食 宜 忌

本茶能減肥瘦身，預防三高。糖尿病可改用不易升血糖的楓糖漿代替；但胃酸過多者、孕婦及婦女來經期間不宜服。

蘋果：有益心臟、提高記憶力、養肝解毒、提升免疫系統。成熟的蘋果甜美清香，但個頭超乎平常太大，顏色過分紅艷，太過油光者不宜購買。

田七花茶

材料（2 人量）
· 田七花 8-10 朵

做法

1　田七花放入壺內，用開水
　　沖洗一遍，將水倒走。
2　再注入 500 毫升開水，焗
　　7 分鐘可飲用。

田七花：具清熱、平肝、降壓等
功效。但人參花和田七花外形
極相似，人參花療效遠不如田七
花，售價亦平，故坊間有不良
商販會以人參花假冒田七花出
售。選購時以花朵大而花蕾密
集、不帶柄或帶短柄、氣清香、
質脆易碎、乾燥無蟲蛀者為佳。

食 療 功 效

提神補氣、預防三高。對
肥胖、三高症、精神不
振、頭昏、耳鳴、失眠、
急性咽炎及血瘀體質者有
幫助。

飲 食 宜 忌

本品味甘微苦，有點人參
味道，可預防三高症。但
孕婦及婦女來經期間忌
服。

杜仲葉
炒黑豆茶

材料（2人量）
- 杜仲葉 10 克
- 炒香青仁黑豆 50 克

做法

1　杜仲葉、炒黑豆放入壺內，用開水沖洗一遍，將水倒走。

2　再注入 500 毫升開水，焗 15 分鐘可飲用。

食療功效

補肝益腎、消脂降壓。適合肥胖、三高症、失眠多夢、腎虛腰痛、皮膚粗糙暗淡及陽虛、氣虛者服用。

飲食宜忌

本品味甘香，頤神養性，是腎虛者佳品，孕婦可服。但黑豆炒後，熱性較大，多食易上火，陰虛火旺者及痛風人士不宜服。

認識主料

杜仲葉：可以補肝腎、強筋骨、抗衰老。選購以墨綠色偏黑，將葉撕開有白色細密的拉絲，葉片小，有草腥氣味，乾燥、無生蟲者為佳。

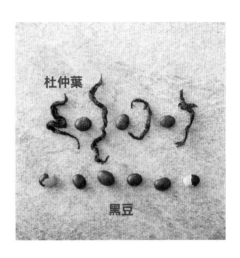

黑豆：青仁黑豆肉青色、皮黑色，含有的花青素能清除體內自由基，尤其是在胃酸的環境下，抗氧化效果好，常食能延緩衰老。中醫認為青色入肝、黑色入腎，故青仁黑豆能肝腎雙補。

五 味 降 壓 湯

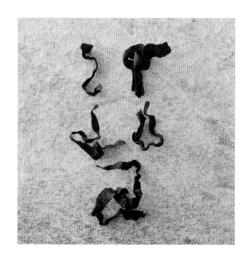

認 識 主 料

螺旋海藻：呈疏鬆或緊密的螺旋形彎曲，故而得名。具有減輕癌症放療、化療的毒副作用、提高免疫力、抗輻射、預防三高症等功效；以產自非洲、墨西哥者較優質。

材料（3～4人量）

- 芹菜 60 克
- 番茄 200 克
- 螺旋海藻 3 克
- 馬蹄 8 粒
- 洋葱 1 個

調味

- 海鹽半茶匙

做法

1 芹菜洗淨，切粗粒；洋葱、馬蹄、番茄洗淨去皮，切粗粒；螺旋海藻沖洗。

2 燒熱 800 毫升水，加入全部材料，煮 20 分鐘，調味即成。

食療功效

清熱排毒、消脂降壓。適合高血壓、高血脂、高膽固醇、高血糖、食慾欠佳及痰濕體質者食用。

飲食宜忌

本品清香味美，能延緩衰老，調整身體酸鹼值，防癌抗癌；但脾胃虛寒及陽虛者慎服。

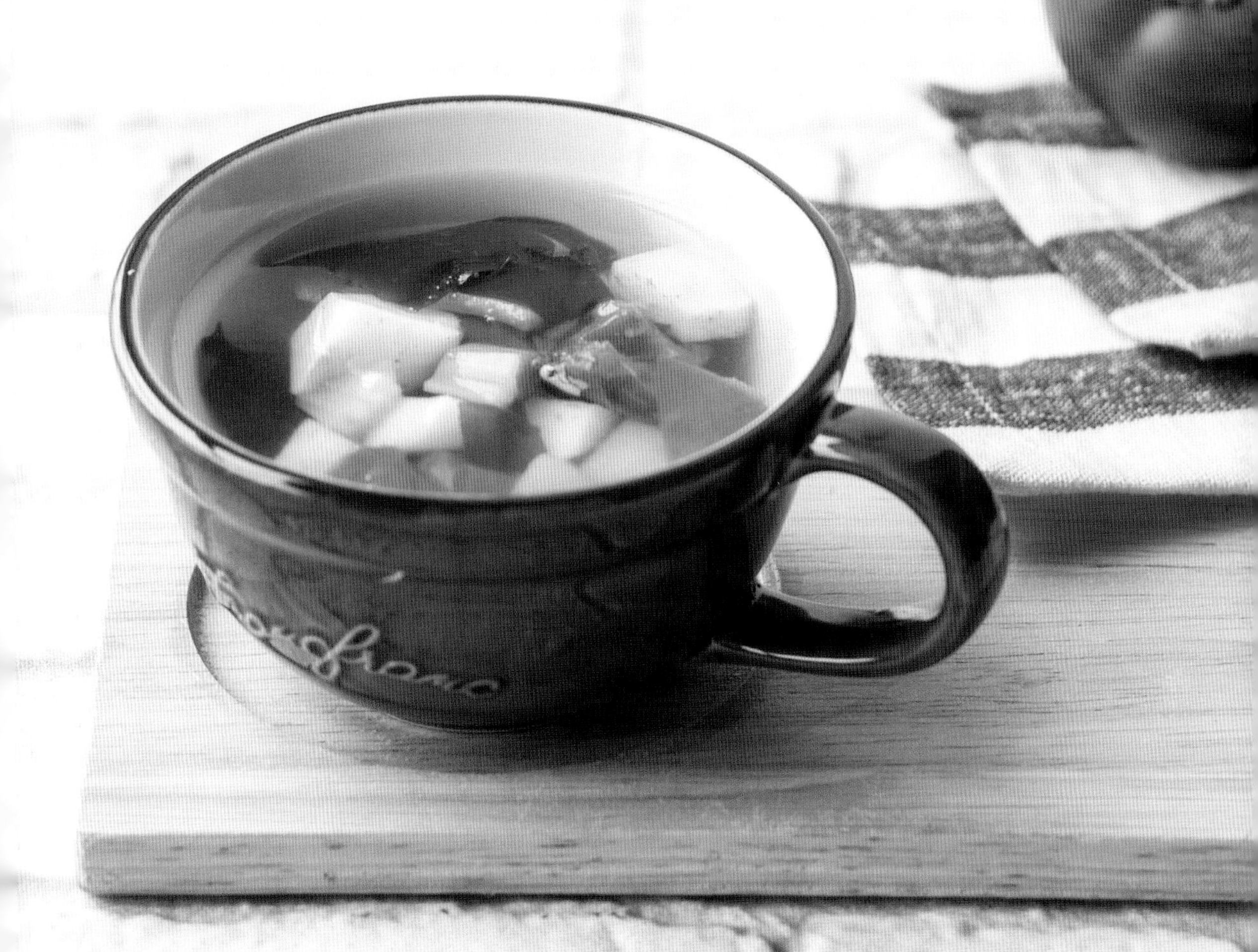

香菇瑤柱
豆腐羹

材料（3～4 人量）

- 香菇 30 克
- 瑤柱 3 粒
- 豆腐 1 磚
- 青豆粒 1 湯匙
- 上湯 800 毫升

調 味

- 海鹽半茶匙
- 胡椒粉半茶匙

芡 汁

- 生粉 1 湯匙
- 水 1 湯匙

做 法

1 香菇、瑤柱浸軟；香菇切
   幼絲，瑤柱撕成細絲。
2 豆腐沖洗後切細絲；青豆
   粒解凍後出水。
3 燒熱上湯，加入香菇絲、
   瑤柱絲煮 20 分鐘，再加
   入豆腐絲及青豆粒，下調
   味，滾起加入芡汁，邊煮
   邊攪，汁濃稠即成。

認 識 主 料

瑤柱：又稱乾貝，有滋陰養血、
健脾補腎作用。優質瑤柱呈淡黃
色，有白霜者味濃；顏色啡黑者
存放太久，味道欠佳。

瑤柱

香菇

香菇：素有「植物皇后」之美
譽。能消脂降壓、提高免疫力、
抑制腫瘤、抗衰老、抗輻射。選
購以菇肉白色，厚身完整，具香
味，菇柄短小者為佳。

食 療 功 效

補中益氣、清熱生津。適合肺
腎陰虛、精力不足、高血壓、
糖尿病、飲食積滯及氣虛、陰
虛者食用。

飲 食 宜 忌

本品鮮味可口，營養豐富，老
少皆宜；但痛風症患者不宜食。

材料（3～4人量）
- 南瓜 250 克
- 蒜子 50 克
- 田雞 2 隻
  （約 300 克）
- 生薑絲 1 湯匙

調味
- 海鹽半茶匙

做法

1. 南瓜連皮洗淨，切塊；蒜子去衣；田雞劏洗淨，斬件後出水。

2. 將材料放入煲內，用 1.5 公升水煮至大滾，改用文火煮 1 小時，調味即成。

南瓜蒜子田雞湯

南瓜：含豐富果膠，果膠有很好的吸附性，能黏結和消除體內細菌毒素和其他有害物質，如重金屬中的鉛、汞和放射性元素，起到解毒作用。所含微量元素鈷，能刺激胰島素生長，對糖尿病者有益。糖尿病者宜揀長身味道較淡的品種食用。

食 療 功 效

滋陰補虛、利水解毒。適合胃病、糖尿病、高血壓、妊娠水腫及氣陰兩虛者服用。

飲 食 宜 忌

本品清甜可口，營養豐富，能保護胃腸黏膜，幫助排走身體重金屬等毒素，老少皆宜；但濕重者不宜多食。

蒜子：除了有殺菌消炎作用，還可抗動脈粥樣硬化、降血壓、降血脂及抗腫瘤作用。蒜子配南瓜，能提高人體對葡萄糖的耐量，糖尿病者服食，可降低血糖含量。

田雞：因肉質細嫩勝似雞肉，故稱田雞。有護心、通乳、養肝、提高免疫力、健腦、壯骨、養陰補虛、抗輻射等功效。但田雞體內多含有寄生蟲，必須煮到熟透方可食用。

豬橫脷：是豬的胰臟，能強肝健胃、幫助消化。豬橫脷如後下，湯色很清很好味，如煮的時間過長，湯會呈烏黑色。

粟米鬚：新鮮淺黃色味清香，陳久之後呈褐色，兩者功效相若，有利尿消水腫、消脂、降血糖功效。乾品用 10 克即可。

食 療 功 效

健脾補腎、預防糖尿。適合動脈硬化、高血糖、高血脂、肥胖、水腫、脾虛泄瀉及氣虛者服用。

飲 食 宜 忌

本品能有效降血糖，糖尿病者可常服用。任何體質人士可服，但有外感發燒者慎服。

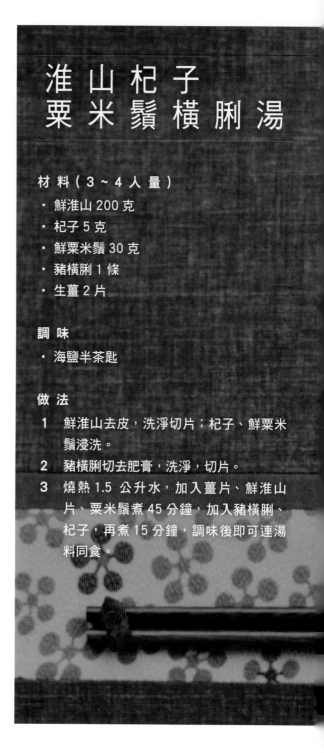

淮山杞子
粟米鬚橫脷湯

材 料（ 3 ～ 4 人 量 ）
- 鮮淮山 200 克
- 杞子 5 克
- 鮮粟米鬚 30 克
- 豬橫脷 1 條
- 生薑 2 片

調 味
- 海鹽半茶匙

做 法
1 鮮淮山去皮，洗淨切片；杞子、鮮粟米鬚浸洗。
2 豬橫脷切去肥膏，洗淨，切片。
3 燒熱 1.5 公升水，加入薑片、鮮淮山片、粟米鬚煮 45 分鐘，加入豬橫脷、杞子，再煮 15 分鐘，調味後即可連湯料同食。

粉葛眉豆薏米排骨湯

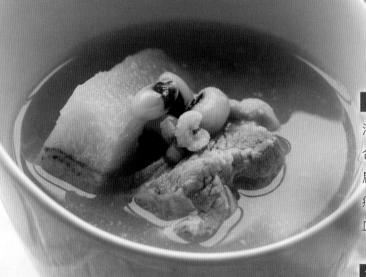

食療功效

清熱下火、預防三高。適合經常飲酒、頭暈頭痛、肩頸肌肉緊張、上火喉痛、二便不利、三高症及血瘀體質人士服用。

飲食宜忌

本品清甜味美,三高症及飲酒過度者可常服;但脾胃虛寒者慎服。

認 識 主 料

粉葛：性味甘涼，可治感冒頭痛、高血壓、頸項強痛，並能解熱除煩、生津止渴、解酒毒、降壓、減肥。選購以塊根肥大，切面粗糙，質硬而重，粉性足，纖維性少者為佳。

材料（3～4人量）

- 粉葛 300 克
- 眉豆 50 克
- 薏米 30 克
- 紅棗 6 粒
- 陳皮 1 塊
- 排骨 300 克

調味

- 海鹽半茶匙

做法

1 粉葛去皮，洗淨切塊；眉豆、薏米、陳皮浸洗；紅棗去核。
2 排骨洗淨，出水。
3 燒熱 1.8 公升水，放入全部材料，先用大火煮滾，改用文火煮2 小時，調味即成。

眉豆：含易於消化的優質蛋白，所含的磷脂有助促進胰島素分泌，是糖尿病人的理想食品；但氣滯便結者宜少食。

預防低血壓的食療

血壓下降的程度低於正常值，且維持一段時間，即為低血壓。一般成年人若血壓測量多次，都維持收縮壓在 90 至 100mm/Hg 以下，而舒張壓在 50 至 60mm/Hg 以下時，就應接受更進一步的身體檢查，找出原因而加以治療。

低血壓雖然不算是一種疾病，但可能是其他疾病所致，而且它會使人頭暈眼花、精神疲憊、注意力不集中或昏倒、休克，而導致其他傷害產生。低血壓病因不明，醫學研究指出，百分之三至七的低血壓患者屬於本態性低血壓，這類病人的心臟收縮能力和血管的抵抗是正常的；而繼發性低血壓則有明顯的疾病致因，例如心臟疾病、循環障礙引起低血壓；因細菌或毒素侵犯末稍血管引致；亦有因手術、外傷的大量出血，引致暫時性低血壓；甲狀腺機能低下，甲減患者（Hypothyroidism）亦可能出現內分泌障礙性低血壓。

打工仔女要避免低血壓必須要勞逸結合，日常要有充分休息，同時要注意多做運動。本身有低血壓症狀者切忌捱夜，要有足夠睡眠，多洗熱水澡，加强血液循環；不要久站或突然改變姿勢；不宜在悶熱的環境中工作太久，同時不要穿太緊或高領衣服，因為容易引致血壓驟降而暈倒。

血壓偏低者若營養不足將使血壓更低；若加強營養則可使血壓接近正常值，伴隨的不適症狀也可減緩或消失。日常宜多吃生薑、桂圓、紅棗、核桃、人參、五味子、山藥、百合、蜂蜜等滋補的食物，有助改善低血壓現象。

低血壓不是很嚴重的病症，所以只要平時飲食上多注意，在食物的選擇上，多吃一些可以升高血壓的食物，少吃或不吃降低血壓的食物，如芹菜、西瓜、冬瓜、苦瓜、綠豆、海帶、蘿蔔等，慢慢血壓就會達到正常值。

認 識 主 料

當歸

一般生用，為加強活血功效可用酒炒製。當歸頭補血，歸身活血，歸尾袪瘀，止血則用當歸炭；產婦宜用全當歸。

黃蓍

能補氣升陽、益衛固表、利水消腫、托毒生肌。由於產於內蒙古、甘肅、黑龍江一帶的北方原野，故又稱北蓍。選購以粗細均勻、皺紋少、斷面色黃白、味香甜者為佳。

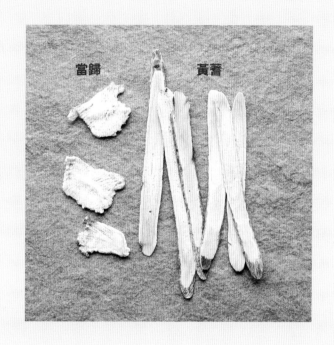

當歸　　　黃蓍

當 歸 黃 蓍 雞 蛋 茶

材料（3～4 人量）

- 當歸頭切片 12 克
- 黃蓍 20 克
- 杞子 5 克
- 雞蛋 2 個

調味

- 紅糖 1 湯匙

做法

1　當歸、黃蓍、杞子浸洗；雞蛋焓熟，去殼。
2　將當歸、黃蓍用 700 毫升水煮 30 分鐘，加入焓熟雞蛋、杞子及紅糖，煮至糖溶即可供食。

食療功效

補益氣血、提升血壓。適合氣虛血弱、氣短乏力、面色萎黃、唇色蒼白、頭暈目眩、心悸肢麻者飲用。

飲食宜忌

本品補益氣血，婦女及血虛者可常服；但陰虛火旺、外感發燒、痰濕者及有出血症者忌服。

黨參黃精茶

材料（2 人量）

- 黨參 20 克
- 黃精 20 克
- 炙甘草 6 克
- 紅棗 8 粒

做 法

1 材料浸洗，黨參切薄片，紅棗去核。

2 將全部材料用 800 毫升水煮半小時即可飲用。

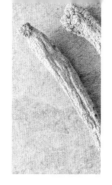

認 識 主 料

黨參：益氣、生津、養血，以產自山西上黨區最著名。選購以切面黃白色或黃棕色，有菊花紋，中央有淡黃色圓心，皺紋清晰、氣香無蟲蛀，味微甜者為佳。

食 療 功 效

健脾益氣、提升血壓。適合氣血虛弱、頭暈氣短、心悸、自汗、失眠多夢、神經衰弱、貧血者飲用。

飲 食 宜 忌

本品甘甜，無論氣津兩傷或氣血雙虧的低血壓者都有益；但脾虛泄瀉及痰濕氣滯者不宜服。

黃精：功能補脾潤肺，養陰生津。中藥店售賣的多為反覆蒸製的熟黃精，以切片油潤、無雜質、無蟲蛀者為佳。

參鬚圓肉紅棗茶

材料（2 人量）

- 紅參鬚 10 克
- 圓肉 10 粒
- 紅棗 6 粒

做法

1. 紅棗去核切片，紅參鬚、圓肉同放入壺內，用開水沖洗一遍，將水倒走。
2. 再注入 500 毫升開水，焗 15 分鐘可飲用。

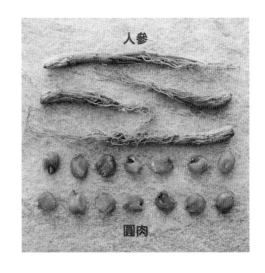

認識主料

人參：共分三個部分：蘆頭、參身、參鬚。蘆頭主要是解熱、催吐藥材；參身是補氣強身，提升免疫力的補益品；參鬚屬涼補之品，補氣之力較微弱但不易上火，適合泡茶服。而參鬚因加工方法不同分紅直鬚、白直鬚、紅彎鬚、白彎鬚品種。泡茶以直參鬚為佳。

圓肉：含有多種營養物質，有補血安神、健腦益智、補養心脾的功效。最適合貧血、低血壓者食用，但懷孕婦女不宜過量食圓肉，因易上火。生曬圓肉以色澤黃亮的質好，深黃帶紅褐色的質次；烘焙圓肉色澤深黃帶紅的較好，紅褐帶黑的為次。

食療功效

健脾益氣、補血安神；適合氣血虛弱、面色蒼白、昏昏欲睡、消化不良、容易頭暈眼花者飲用。

飲食宜忌

本品香甜味美，益氣生津，男女可服。但有外感發燒、便秘有實熱症者忌服。服用本茶，忌吃蘿蔔及飲用濃茶、咖啡。

人參黃蓍
五味牛膔湯

材料（3～4人量）

- 人參 5 克
- 黃蓍 15 克
- 五味子 10 克
- 圓肉 8 粒
- 金錢膔 1 條（約 250 克）

調味

- 海鹽半茶匙

做法

1 金錢膔切塊，出水；人參、黃蓍、五味子、圓肉浸洗。
2 將全部材料放入燉盅內，注入 700 毫升開水，隔水燉 3 小時，調味即可供食。

補氣滋陰、強壯體質；適合氣血兩虛、手術後或大病後面色蒼白、神疲乏力、自汗、盜汗、難以入睡者服用。

五味子

人參

認識主料

人參： 大補氣血，紅參是經蒸煮過，再曬乾的人參，其補益功效較白參強，較適合陽虛、心力不足者；陰虛內熱體質宜用白參。紅參以顏色紅潤，氣味濃香，質硬而脆者為佳。

五味子： 五味俱全，可以保護人體五臟——心、肝、脾、肺、腎。五味子具有消炎作用，能防治肝臟損傷，是病毒性肝炎患者的良藥。但有些人服五味子後有「火燒心」的感覺，故最好配搭一些健胃藥一齊服用。近年五味子常用於治療神經衰弱症。

飲食宜忌

本品清香味美，對貧血、低血壓，及手術後人士最有幫助；但外感發熱，陰虛內熱者及懷孕期間忌服。

淮山栗子核桃圓肉雞湯

材料（3～4人量）

- 淮山 60 克
- 栗子 100 克
- 核桃肉 80 克
- 圓肉 10 克
- 紅棗 6 粒
- 走地雞 1 隻
- 生薑 3 片

調味

- 海鹽半茶匙

做法

1　走地雞劏洗淨，斬件後出水；栗子投入開水中汆水，去衣。

2　紅棗去核；淮山浸洗；圓肉沖洗。

3　將全部材料放入煲內，用 1.7 公升水煮至大滾，改用文火煮 2 小時，調味即成。

栗子： 是一種香甜佳果，有「腎之果」的美譽。能養胃健脾、補腎強筋。以外形玲瓏，果殼有光澤，果肉呈米黃色，糯性強，甘甜芳香者為佳。

食 療 功 效

益氣養血、健脾補腎。適合腎虛、血壓低、腰膝酸軟、兩腳無力、自汗怕風、頭昏耳鳴、記憶力減退及陽虛體質者服用。

飲 食 宜 忌

本品清香味美，健腦強身；但外感發燒、陰虛火旺及脾胃虛弱者忌服。

核桃： 温肺補腎，能增強記憶，但性温多油脂，過食易生熱聚痰。選購以核桃個頭要均勻，殼上的凹痕越大越粗糙的表示核桃殼越薄，反之殼就越厚，最厚的「鐵核桃」不易敲開；外殼發黑、泛油的多數為壞果。

預 防 眼 睛 疲 勞 的 食 療

倘若工作需全神貫注看電腦螢幕，長期對着電腦工作、眼睛眨眼次數減少，就會造成眼淚分泌相應減少，同時閃爍螢屏強烈刺激雙目，引致眼睛疲勞。有些打工仔由於習慣熬夜，玩手機、看電視，長期睡眠不足，亦會引致視疲勞。患者一般的症狀是視物稍久即模糊，甚至出現眼睛乾澀、頸、肩、頭等部位出現疼痛、頭昏腦脹，嚴重時可能出現噁心、嘔吐等症狀。

對於眼睛的保健，除了多接近大自然，多看遠處及綠色植物，日常生活中也要讓眼睛有足夠的休息時間。此外，工作時間要多飲水，讓眼睛稍作休息及有水分滋潤，飲食中更需要適當補充維生素及礦物質。

合理補充眼睛所需的營養素，如維生素 A 的食物，對保護眼睛、防止視力傷害、提高視力非常重要。含維生素A的食物如紅蘿蔔、南瓜、甜薯、菠菜、藍莓、海藻、魚肝油、動物肝臟等日常可以多食；眼睛疲勞者要注意飲食上營養均衡，平時多吃些粗糧、雜糧、紅綠色蔬菜、豆類、水果等含有維生素、蛋白質和纖維素的食物亦有助改善眼睛疲勞。

認 識 主 料

桑椹子

可改善皮膚（包括頭皮）的血液
供應，營養肌膚及黑髮，並有明
目、提神、補血、生津、潤腸等
功效。以紫黑色粒大厚實者療
效佳，未成熟的桑椹子含氰胺
酸，不宜食。

黑杞子

含豐富花青素，能滋補肝腎、
興奮大腦神經、抗衰老及美容
等。以顆粒大、大小平均、果實
連果柄、乾燥無霉變、泡茶呈漂
亮的紫藍色者為佳。

桑椹子

黑杞子

黑 杞 子 桑 椹 茶

食療功效

滋陰養血、益肝明目。適合肝
腎陰虛、兩眼昏花、眼睛疲勞
乾澀、頭暈目眩、頭髮斑白、
腰膝酸軟、腸燥便秘者服用。

材料（2 人量）

· 黑杞子 10 克
· 桑椹子 30 克

調味

· 蜂蜜 1 湯匙

做法

1 黑杞子、桑椹子放入
壺內，先用開水沖洗
一遍，將水倒走。
2 再注入 500 毫升開
水，焗 10 分鐘，調
入蜂蜜飲用。

飲食宜忌

本品甜酸可口，養顏黑髮，延
緩衰老，一般人可服。但脾虛
泄瀉者忌服，孕婦慎服。

夏枯草金銀花茶

材 料（2～3人量）

- 夏枯草 30 克
- 金銀花 10 克

調 味

- 冰糖 30 克

做 法

1　夏枯草、金銀花洗淨。
2　用 1 公升水先將夏枯草煮 45 分鐘，加入金銀花、冰糖，煮 5 分鐘即成。

夏枯草： 能清肝明目、清熱瀉火。但飲用前最好了解一下飲用者體質，因夏枯草性寒涼，對熱性體質及體力勞動者較適合。

食 療 功 效

清肝明目、消炎去翳。適合肝火上炎、目赤腫痛、眼眵多、急性咽炎、頭痛、眩暈及氣鬱體質者飲用。

飲 食 宜 忌

本品為家常涼茶，男女可服；但氣虛、濕重、脾胃虛弱者常服，易造成腹瀉甚至加重症狀；孕婦及婦女來經期間忌服。

金銀花： 冬天不葉落，故名「忍冬」。花初開時為白色，後變為金黃色，白色、黃色的花會同時長在枝藤上，故又稱金銀花。金銀花以花蕾入藥，以乾燥、氣香、無蟲蛀者為佳；六磷酸葡萄糖去氫酵素缺乏症（俗稱蠶豆症 /G6PD）人士忌用金銀花。

決明菊花茶

材料（2 人量）

- 炒決明子 20 克
- 菊花 6 克

做法

1 炒決明子、菊花放入壺內，先用開水沖洗一遍，將水倒走。
2 再注入 500 毫升開水，焗 10 分鐘可飲用。

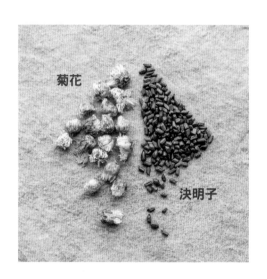

菊花

決明子

認識主料

菊花：可選用杭菊、貢菊、胎菊。以花蕾緊密、乾燥、氣香、無生蟲、無霉變者為佳。

決明子：必須用白鑊炒至微有香氣才能泡茶，否則不易出味。決明子含有一種大黃素類似瀉藥成分，不宜長期飲用，以免損傷身體正氣。

食療功效

清肝明目、抗菌消炎。適合肝火上擾，風熱上壅頭目引致的目赤腫痛、眼乾多淚、頭痛眩暈、目暗不明及陰虛體質者飲用。

飲食宜忌

本品味甘微苦，對肝炎及高血壓人士亦適合；但脾胃虛寒、脾虛泄瀉、低血壓及孕婦不宜服。

淮杞珍珠肉湯

材 料（3～4 人 量）

· 淮山 50 克
· 杞子 6 克
· 紅蘿蔔 1 條
· 生薑 2 片
· 珍珠肉 80 克
· 雞胸肉 1 塊

調 味

· 海鹽半茶匙

認 識 主 料

珍珠肉：是珍珠蚌的
肉，經天然生曬而成。珍
珠肉有清熱解毒、收斂
生肌、滋養肝腎、明目
安神功效。產自澳洲、
美國的較優質，以蚌肉
晶瑩通透、色帶微紅光
澤、氣香及乾燥者為佳。

做 法

1 　淮山、杞子、珍珠
　　肉浸洗；紅蘿蔔去
　　皮，切塊；雞胸肉
　　切片，與珍珠肉齊
　　出水。

2 　將全部材料放入燉
　　盅內，注入 700 毫
　　升開水，隔水燉 3
　　小時，調味即可供
　　食。

滋養肝腎、明目消翳。適合工
作過勞、肝腎陰虛、眼睛疲勞、
視力衰退、夜尿多者服用。

本品清甜美味，補而不燥，適
合眼睛疲勞者飲用；但外感發
燒及痛風者不宜。

夜香花杞子雞肝湯

材料（3~4人量）

- 夜香花 60 克
- 杞子 10 克
- 雞肝 3 ～ 4 副
- 生薑絲 1 湯匙

調味

- 海鹽半茶匙

做法

1　夜香花、杞子洗淨；雞肝洗淨，切開。
2　將雞肝、薑絲用 800 毫升水煮 20 分鐘，加入杞子、夜香花，再滾 5 分鐘，調味即成。

認識主料

雞肝：是補血、養生佳品，可防止眼睛乾澀、疲勞；以色紫紅細嫩者為佳，啡黃色雞肝其實屬病態的「脂肪肝」，味甘香但不宜多食。

夜香花：能清肝明目、去翳、拔毒生肌、祛風除濕。花味清香，以花蕾顆粒大，色綠青翠，含苞未開放者香氣濃。

食療功效

疏風清熱、明目退翳。適合長期對着電腦工作、經常玩手機、看電視、視力疲勞者服用。

飲食宜忌

本品清香味美，能明目退翳，對筋骨酸軟疼痛都有幫助。一般體質皆可服，但痛風患者宜用瘦肉代替雞肝。

預 防 失 眠 的 食 療

很多打工仔由於工作、家庭或朋友間的壓力，壓得喘不過氣來，因此晚上思慮過度造成失眠。初患失眠的表現是入睡困難，不能熟睡；有些則入睡後容易被驚醒，有對聲音敏感，有對燈光敏感，亦有因頻頻從噩夢中驚醒，醒後難以再入睡。

大部分打工仔在經歷到壓力、刺激、興奮、焦慮時；或生病時；或輪班工作睡眠規律改變時，都可能會有短暫性失眠障礙。這類失眠一般會隨着事件的消失或時間的拉長而改善，但是短暫性失眠如處理不當，部分人會導致慢性失眠。

慢性失眠患者每晚睡前就開始胡思亂想，開始擔心睡不好，這種擔心和焦慮及情緒張力增高，反而更加睡不着。因此形成惡性循環，睡不着就更擔心，擔心更睡不着，以致長期失眠。長時間的失眠不但會導致神經衰弱和抑鬱症，並且會嚴重影響肝臟功能。因為晚上 11 時至凌晨 1 時是肝排毒的時間，如果在這段時間熟睡，整個肝臟被肝血所浸潤，就能夠順利將體內毒素過濾排走，如果此時不睡覺，肝的排毒運作受阻，長期失眠自然容易傷肝。

失眠患者要留意不可暴飲暴食，宜少量多餐，穩定血糖；戒宵夜，讓胃腸休息，這樣才較易入睡。睡前可以飲杯暖奶或喝點熱甘菊茶助安眠。平時多補充含色胺酸的食品及維生素 B 群，如核桃、全麥類、豆類、牛奶、乳酪、雞蛋、小麥胚芽、動物肝臟、魚肉等，幫助大腦釋放更多五羥色胺物質，更有助安眠。

失眠並非嚴重疾病，故患者不要過於緊張，只要改善作息時間，睡前盡量少用電腦，不玩網上遊戲，調節好睡眠週期，配合食療，熱水泡腳等方法，失眠症狀應該可以得到改善。

認 識 主 料

百合

所含的秋水仙鹼能抑制癌細胞及
治急性痛風，也是潤肺止咳、清
心安神佳品。以產自湖南等地之
龍牙百合最有名。曬乾的百合以
片長、肉厚、心實、色呈黃白色
或稍帶粉紅色、氣香味不苦者為
佳。

麥米

性味甘涼，能養心安神、除煩、
止汗。選購以顆粒大小均勻、氣
味清香、乾燥無生蟲者為佳。

麥米

百合

百 合 麥 米 蓮 芯
安 神 茶

材 料（ 2 ~ 3 人 量 ）

· 百合 20 克
· 麥米 30 克
· 蓮芯 5 克
· 紅棗 6 粒

調 味

· 冰糖 50 克

做 法

1　百合、麥米浸洗；蓮芯沖洗；紅棗去核。

2　將全部材料放入煲內，用 1 公升水煮至大滾，改用中慢火煮 1 小時，加入冰糖煮溶即可供服。

食 療 功 效

益氣養陰、清心安神。適合陰虛、氣鬱體質，出現精神恍惚、心情抑鬱、婦女臟躁、煩躁不安、失眠、自汗、盜汗及小便不利者服。

飲 食 宜 忌

本品香甜可口，老幼可服；但風寒咳嗽、虛寒出血、外感發燒者不宜。

洋甘菊茶

材料（2人量）

· 洋甘菊 10 克

做法

1　洋甘菊放入壺內，用
　　開水沖洗後將水倒走。

2　再注入 500 毫升開
　　水，焗 5 分鐘可飲用。

食療功效

能舒緩壓力、幫助睡
眠。適合壓力大、情緒
焦躁、失眠多夢、眼睛疲
勞、口氣大及氣鬱體質者
飲用。

飲食宜忌

本品清香可口,常服能潤澤肌
膚、改善女性經前不適、促進睡
眠,但孕婦不宜。

認識主料

洋甘菊:是助眠佳品,最好在飯
後或睡前一小時開始飲,讓身體
自然調整到睡眠狀態,另一方
面亦能幫助消化。配檸檬馬鞭草
同泡,舒壓及舒眠效果更佳。甘
菊以羅馬甘菊、德國甘菊質量較
優,宜購買有機產品,避免農藥
殘留問題。

炒棗仁
柏子仁茶

材料（2 人量）

- 炒棗仁 15 克
- 柏子仁 15 克

調味

- 蜂蜜 1 湯匙

做法

1. 炒棗仁、柏子仁放入壺內，用開水沖洗後，將水倒走。
2. 再注入 500 毫升開水，焗 10 分鐘，調入蜂蜜可服。

食療功效

養心安神、治療失眠。適合陰虛血虛、神經衰弱、失眠多夢、盜汗、腸燥便秘者服用。

飲食宜忌

本品甘香微酸，為失眠及陰血虧虛者之佳品，產後失眠最適宜；但孕婦及痰濕體質者不宜。

認識主料

生酸棗仁：炒至微香用，適合泡茶飲用，較易出味。以粒大飽滿、有光澤、外皮紅棕色、種仁黃白色為佳。

生酸棗仁

柏子仁

柏子仁：是側柏的種仁，是養心安神、潤腸通便佳品。但易走油變化，不宜曝曬，宜放雪櫃或置陰涼處存放。以氣香、色澤鮮明、無出現「哈喇味」（油脫味）者為佳。

靈芝圓肉茶

材料（2人量）

· 黑靈芝片 20 克
· 圓肉 10 粒

做法

1　黑靈芝片、圓肉放入壺內，用開水沖洗，將水倒走。
2　再注入 500 毫升開水，焗 15 分鐘即可飲用。

食療功效

補血安神、幫助睡眠。適合氣血虛弱、心神不寧、失眠多夢、驚悸、精神不振、甲亢、便溏腹瀉者服用。

飲食宜忌

本品味甘微苦，有良好補腦安神作用。但手術後、剖腹產後不宜多服，以免出血量多。外感發燒者忌服。

認識主料

靈芝：古代按靈芝的六色來區分療效，青芝補肝明目、安神、增強記憶；赤芝補中益氣、疏肝解鬱、強壯體質；黃芝益脾胃、安神；白芝止咳益肺、安神；黑芝利水道、益腎氣、安神；紫芝益精氣、強筋骨、安神。當中以赤芝療效最佳，但最苦；黑芝味不太苦，可作泡茶用。

人參麥冬茯神茶

白參：是以水參為原料，將之剝皮並自然曬乾後，會呈現微白的黃色，因此稱為白參。性平和，但功效較弱，適合做一般強身健體的保養藥材之用，對失眠而有虛熱者較適合。

材 料（2 人 量）

- 白參片 5 克
- 麥冬 10 克
- 茯神 15 克
- 圓肉 8 粒

調 味

- 冰糖 30 克

做 法

1 麥冬、茯神浸洗；白參片、圓肉沖洗。

2 將麥冬、茯神、白參片、圓肉放入煲內，用 800 毫升水煮 1 小時，加入冰糖煮溶即成。

茯神：有多種，有些白色一片片，有些粒狀，有些是抱住松根而生的野生茯神，中間有粒「眼」就是松樹的根。茯苓菌很大，其啡色表皮稱茯苓皮，有利水功效；白色肉卷狀稱雲苓，能健脾祛濕；中間部分的才稱茯神，有良好安神作用。

益氣養血、寧心安神。適合用
腦過度、精神不振、失眠多夢、
口乾舌燥及氣陰兩虛者服用。

飲 食 宜 忌

本品參味香濃，對工作過勞及
長期失眠者有益。但感冒發熱
者、孕婦及小孩不宜。服用人
參不宜飲濃茶及吃白蘿蔔，以
免抵消藥效。

百合杞子鮑魚湯

材料（3～4 人量）

- 百合 50 克
- 雪耳 5 克
- 杞子 6 克
- 生薑 2 片
- 鮮鮑 300 克

調味

- 海鹽半茶匙
-

做法

1 百合、杞子浸洗；雪耳浸軟，去蒂；鮮鮑去腸臟，連殼出水備用。

2 將全部材料放入煲內，用 1.5 公升水煮至大滾，改用文火煮 2 小時，調味即可飲用。

認識主料

百合：能清心除煩、寧心安神。選購以乾燥、色白微黃有光澤、肥大均勻、少碎屑、無蟲蛀、無焦片及嫩芯者為佳。

鮑魚：具養血柔肝、滋陰清熱、益精明目功能。鮑魚殼是中藥石決明，治療頭痛眩暈、視物昏花，故鮮鮑宜連殼一齊用。鮮鮑以個體均勻、橢圓形、體潔淨、背面凸起、肉厚、色澤鮮明者為佳。

夜交藤：是何首烏的藤莖，因在晚上藤會自動相互交合，故名。夜交藤有養心安神、祛風活絡功效。以枝條粗壯，均勻，外皮棕紅色者為佳。

夜 交 藤
瑤 柱 瘦 肉 湯

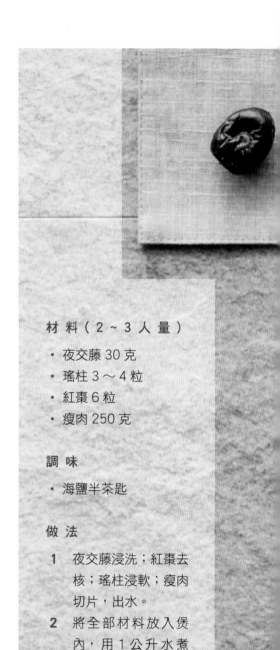

材 料（ 2 ～ 3 人 量 ）

- 夜交藤 30 克
- 瑤柱 3 ～ 4 粒
- 紅棗 6 粒
- 瘦肉 250 克

調 味

- 海鹽半茶匙

做 法

1 夜交藤浸洗；紅棗去核；瑤柱浸軟；瘦肉切片，出水。

2 將全部材料放入煲內，用 1 公升水煮至大滾，改用文火煮個半小時，調味即成。

食 療 功 效

滋陰補腎、養心安神。適合血虛身痛、虛煩失眠多夢、風濕痹痛、肌膚麻木者服用。

飲 食 宜 忌

本品藥味不重，對經常捱夜致虛火上升的不眠者有益；但燥狂實火者忌服。

茯苓遠志燉豬心

食療功效

益肝養血、寧心安神。適
合肝血虛引起的失眠多
夢、健忘驚悸、神志恍
惚、記憶力減退者服用。

飲食宜忌

本品微有藥味，各種體質
有失眠症狀者可服。但有
胃炎及胃潰瘍者慎服，孕
婦忌服。

134

認 識 主 料

遠志：又名遠志通，味辛、苦，有寧心安神、祛痰開竅、解毒消腫作用。以色黃、筒粗、肉厚、乾燥者為佳。

材 料（ 2～3 人 量 ）

- 茯苓 15 克
- 遠志 6 克
- 酸棗仁 10 克
- 圓肉 10 粒
- 陳皮 1 塊
- 豬心 1 個

調 味

海鹽半茶匙

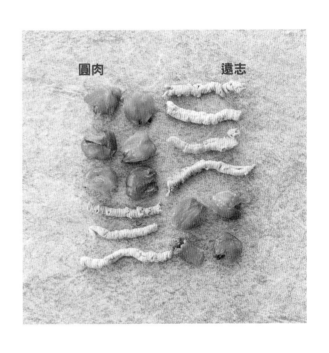

圓肉　　　　　　　　遠志

做 法

1　茯苓、遠志、酸棗仁、陳皮浸洗；圓肉沖洗；豬心切厚片，出水。

2　將全部材料放入燉盅內，注入 600 毫升水，隔水燉 3 小時，調味即成。

圓肉：味甘，性溫，有養血安神、補虛益智等功效。對體虛、記憶力衰退、頭暈失眠者有益；但有上火發炎症狀及懷孕期不宜多食。

預 防 便 秘 的 食 療

便秘的主要徵狀是大便次數減少，間隔時間延長，糞質乾硬，排出困難；或糞質不乾，但排出不暢。常伴有腹脹、腹痛、食慾減退、噯氣反胃等症狀。正常人的大便次數可以是一日兩次至兩日一次。三日以上才一次大便，或大便太硬都可以視作便秘。

急性便秘多由腸梗阻、腸麻痺、急性腹膜炎、腦血管意外、急性心肌梗塞、肛周疼痛性疾病等急性疾病引起；而慢性便秘可能與腸蠕動功能失調有關，也可能與精神因素有關。

緊張、壓力或憂愁、思慮過度的人，會影響傳導排便的神經，並且使胃腸功能低下，造成便秘。除了這些精神因素，打工仔女普遍睡眠不足，身體過度勞累，飲水少、運動少、纖維質食物吸收少，都會引致便秘。

便秘在程度上有輕有重，在時間上可以是暫時的，也可以是長久的。由於引起便秘的原因很多，也很複雜，因此，一旦發生便秘，尤其是比較嚴重的，持續時間較長的便秘，就應及時到醫院檢查，查找引起便秘的原因，以免延誤原發病的診治。患者切勿濫用瀉藥，因為瀉藥的藥用功能為刺激腸道蠕動，長時間大量使用這類藥物，腸道蠕動就會越來越慢，沒有用藥時，排便也就越加困難。因此要盡量避免長時間、大量使用這種藥物。除非由醫師建議，否則使用瀉藥時間不要超過一星期。長時間濫用瀉藥很容易對身體造成傷害，且濫用瀉藥也可能隱藏疾病病灶，導致延誤治療。

改善便秘方法要多飲水，早晨空腹時，飲用一杯250毫升的暖開水，令腸道有水分滋潤，對排便很有幫助。日常要多吃纖維素豐富的蔬果類食物，牛奶、乳酪、西梅、海藻、蒟蒻等都有助排便；太精緻及難以消化的食物，盡量少吃；睡眠要充足，運動要足夠，以散步減壓能幫助身體消化機能，也可減少便秘的發生。

杏仁

有分甜杏仁及苦杏仁，甜杏仁即俗稱的南杏，苦杏仁即北杏。南杏潤肺止咳、美顏潤膚；北杏溫肺定喘、潤腸通便。北杏入藥但有小毒，故一般南、北杏一齊用，南杏 10 粒，北杏只宜用 2 粒。杏仁粉則方便日常沖服。

鮮奶

具補虛損、益肺胃、生津潤腸的功效。牛奶是人體鈣質的最佳來源。但很多人有乳糖不耐症，飲奶會肚瀉、嘔吐或不適，一般兒童和年輕人較容易吸收，老年人不易消化，可以豆奶代替。

杏仁

鮮奶

杏 仁 奶 茶

材料（2 人量）

· 杏仁粉 50 克
· 鮮奶 500 毫升

調味

· 冰糖 20 克

做法

1　杏仁粉用少許鮮奶調成糊
　　狀，加入鮮奶用中慢火煮
　　7 分鐘。

2　加入冰糖，煮至糖溶即成。

食療功效

潤肺養顏、滑腸通便。適
合久病體虛、氣血不足、
營養不良、皮膚乾燥、乾
咳無痰、腸燥便秘者服。

飲食宜忌

本品香滑滋補，任何體質
可服；但對鮮奶過敏及腹
瀉者不宜服。

火麻仁蜜茶

材料（2 人量）

・火麻仁 20 克

調味

・蜂蜜 2 湯匙

做法

1. 火麻仁用白鑊炒香，搗成細末，放入茶包袋。
2. 將茶包袋放入壼內，用 500 毫升開水沖泡；焗 10 分鐘，調入蜂蜜可飲用。

食療功效

活血補虛、潤腸通便。適合血虛津虧、陰虛火旺、腸燥便秘及習慣性便秘者飲用。

飲食宜忌

本品香甜美味，對經常捱夜睡眠不足、虛火上升者最適合。孕婦、脾胃虛弱及陽虛者忌服。

認 識 主 料

火麻仁：是大麻的種籽，有小毒，不能服用過量，亦不宜長期服用。用白鑊炒至微黃色，有香氣就可以用。選購以顆粒大小均勻、乾燥無蟲蛀者為佳。

羅漢果無花果茶

材料（2人量）

- 羅漢果半個
- 無花果 4 粒

做法

1. 羅漢果沖洗後打碎；無花果切碎，白鑊炒至微焦。
2. 將羅漢果、無花果碎放入壺內，用 500 毫升開水沖泡，焗 15 分鐘可服用。

食療功效

清熱潤肺、滑腸通便。適合陰虛火旺、燥咳聲嘶、咽喉腫痛、腸燥便秘、痔瘡者服用。

飲食宜忌

本品清甜滋潤、老少合飲，對婦女產後氣血不足、乳汁稀少亦有幫助；但脾胃虛寒者慎服。

認識主料

無花果：潤肺止咳、清熱潤腸，又有催乳作用。以顆粒大、色澤鮮明、軟硬適中、無蟲蛀、無霉變者為佳；產自地中海一帶者味道較清甜。

羅漢果：為糖尿病、高血壓、高血脂和肥胖症患者的天然甜味劑，以顆粒大、皮色均勻、搖之不響者為佳。

羅漢果

無花果

松子南瓜羹

材料（3～4人量）

- 南瓜 200 克
- 松子仁 30 克
- 牛奶 500 毫升

調味

- 赤砂糖 50 克

芡汁

- 生粉 1 湯匙
- 清水 1 湯匙

做法

1 南瓜洗淨，去皮及瓤，用蒸
   鑊蒸後壓成泥；松子仁用白
   鑊炒香。

2 牛奶煮滾，放入南瓜泥及砂
   糖，攪拌後勾芡，盛入碗
   中，灑入松子仁即可供食。

松子仁：味甘，性微溫，可扶正補虛、
補腎益氣、滑腸通便及美容抗衰老。選
購以顆粒大、氣香、無「哈喇味」（油
腦味）者為佳。

南瓜：有補中益氣、益肝解毒、防治糖
尿等功效，但過量食用易引發黃疸和腳
氣病，痰濕氣滯者宜少食。

南瓜

松子仁

食療功效

補益氣血、潤腸通便；適合陽虛、氣虛
體質、習慣性便秘及老年體虛者服食。

飲食宜忌

本品香甜美味，老少可食；但脾虛腹瀉
及痰濕者慎服。

霸王花杏仁
紅蘿蔔豬䐈湯

材料（3～4 人量）

- 霸王花 50 克
- 南杏 30 克
- 紅蘿蔔 1 條
- 無花果 4 粒
- 豬脹 250 克

調味

- 海鹽半茶匙

做法

1　霸王花、南杏浸洗；紅蘿蔔去皮，切塊；無花果切開；豬脹切塊，出水。

2　將豬脹、南杏、紅蘿蔔、無花果用 1.7 公升水煮至大滾，加入霸王花，改用文火煮 2 小時，調味即成。

食療功效

清熱潤肺、潤腸通便。適合痰濕體質、多痰咳嗽、腸燥便秘、腸胃積熱、習慣性便秘者服用。

飲食宜忌

本品潤燥滑腸，對熱性便秘者適合；但寒性體質及氣虛便秘者不宜多服。

認識主料

霸王花： 又名劍花、量天尺。夏天會有鮮品出售，鮮品太寒涼，雖然清暑解熱之力較強，但煲出的湯水潺潺，一般都用乾品。乾品以色澤金黃帶青、完整、碎片少者為佳。

改 善 臉 色 暗 啞 的 食 療

一個人氣血充盈臉色自然紅潤、精神奕奕，相反氣血不足臉色自然暗啞無光。不少打工仔女都可能由於工作疲勞、睡眠不足、思慮過度、情緒低落、過度曝曬、吸煙、飲酒等各種原因令到膚色變差。而臉色暗啞是很多辦公室女士常遇到的外觀問題，其實這種暗啞臉色，可以説是初期皮膚老化的徵狀，需要及早作護理。

簡單的護膚方法，例如用絲瓜削皮後按摩臉部，絲瓜的汁液有良好的緊膚、補水、美白、去皺功效；煲湯常用的老黃瓜，瓜瓤通常會被棄掉，其實刮下來的瓜瓤有很多汁液，是天然美白去皺佳品，用作外敷最好不過；此外，薏仁粉亦是美白嫩膚佳品，除了可加入熱飲中食用，亦可以加些蜂蜜、涼開水調成稀糊狀，作外敷磨砂膏使用，調好後輕輕按摩臉部，不但能去除死皮，更能美白嫩膚，對去除老人斑(即疣)亦有幫助，這些天然食材都是價廉物美的護膚佳品。

當然要改善臉色暗啞必須調整作息時間，舒緩壓力問題。作息有規律，戒煙控酒，多補充營養物質，多食健脾宜胃、補氣、補血的食物以改善內分泌失調，臉色自然能得到改善。

臉黃膚色暗啞者可多吃性味甘平及溫熱食物，忌吃生冷寒涼食物及冰凍冷飲。食物以溫和細軟為補，忌食辛辣刺激、燒烤、煙燻類食物。家庭常用的中藥性質平和的如黨參、黃薈、淮山、杞子、蓮子、芡實、圓肉、紅棗等一般都可應用，但大補氣血的藥膳最好在中醫師指導下服用。中藥及食療調養需一點一點慢慢改進，不能操之過急。

認 識 主 料

玫瑰

氣香性溫，有行氣解鬱、和血調經等功效。紅玫瑰疏肝解鬱、調經止痛、養顏美容、瘦身。以花色均勻、花蕾緊密、氣香、無蟲蛀、無霉變者為佳。

桃花

能疏通經絡、擴張末梢神經毛細血管，促進皮膚營養和氧分供給，起到改善臉色效果。以氣香、乾燥、無生蟲者為佳。

茉莉

茉莉的香氣可穩定情緒，還可清新口氣，調節內分泌，潤澤膚色。選購以花蕾緊結，較圓而短小，香氣濃者為佳。

活 血 三 花 飲

食療功效

活血養顏、改善臉色。適合肝鬱氣滯、臉色暗啞、胸脇脹痛、婦女月經不調、積滯、便秘者服用。

飲食宜忌

本品氣味清香，能舒緩情緒，疏通經絡，改善血液循環，令皮膚滋潤，清除皮膚上的黃褐斑、雀斑、黑斑；但孕婦及大便溏薄者忌服。

材料（2人量）

- 玫瑰、茉莉、桃花各5克

調味

- 蜂蜜1湯匙

做法

1. 將三花放入壺內，用開水沖洗一遍，將水倒走。
2. 再注入500毫升開水，焗5分鐘，調入蜂蜜即可供服。

洋參蓮子茶

材料（2人量）

· 花旗參 10 克
· 蓮子 50 克

做法

1　蓮子洗淨，用清水浸 1 小時，用 700 毫升水煮 30 分鐘。
2　加入花旗參，熄火焗 10 分鐘可服用。

花旗參片

認 識 主 料

花旗參：能調節免疫功能、促進醣類代謝等功能，但不易上火，體虛貧血及老人、小孩過敏體質者都可用。選購以色白橫紋多、質輕、參味濃郁、乾燥無蟲蛀者為佳。

食 療 功 效

益氣生津、健脾養顏。適合氣虛、體質虛弱、用腦過度、臉色無華、容易傷風感冒、脾虛泄瀉者飲用。

飲 食 宜 忌

本品滋補養顏，任何體質可服；但外感發燒、便秘者不宜服。

原粒花旗參

圓肉紅棗
黑米茶

材料（2 人量）

- 圓肉 20 克
- 紅棗 8 粒
- 黑米 30 克

做法

1. 黑米用白鑊炒至香氣溢出；紅棗去核，切片。
2. 將全部材料放入壺內，用 500 毫升開水沖泡，焗 15 分鐘可飲。可沖泡至淡。

認 識 主 料

黑米：是稻米中的珍貴品種，有「補血米」、「長壽米」之稱。具天然水溶性花青素，故用水浸會有褪色現象。選購以黑色有光澤、米粒大小均勻，不含雜質，有米香味、無蟲蛀、無碎米者為佳。

紅菜頭番茄
紅腰豆牛肉湯

材料（3～4人量）

- 紅菜頭 150 克
- 番茄 150 克
- 紅腰豆 50 克
- 牛肉 250 克

調味

- 海鹽半茶匙

做法

1 紅菜頭去皮，切塊；番茄去皮，切片；紅腰豆浸水 1 小時；牛肉切片，出水。
2 將牛肉、紅腰豆用 1.5 公升水煮 1 小時，加入紅菜頭、番茄，滾 10 分鐘，調味即可供服用。

食療功效

護肝補血、改善臉色。適合面色蒼白或萎黃、眼睛疲勞、毛髮枯乾、骨質疏鬆者服食。

飲食宜忌

本品清甜美味，常服能令面色紅潤，任何體質皆宜；但胃潰瘍、胃氣脹、胃酸過多者慎服。

認識主料

紅菜頭、番茄：都屬抗氧化、防癌、抗衰老、抗自由基的食物，常食有利平衡因肉食過量的酸性體質，對預防骨質疏鬆亦有幫助。紅菜頭生食療效更佳；而番茄必須選擇熟透了的，青色未熟的番茄含有毒龍葵素，不宜食用，要擺放至轉了紅色方可食。

番茄

紅菜頭

蟲草花雪耳杷子豬膶湯

材料（3~4人量）

- 蟲草花 30 克
- 雪耳 15 克
- 杞子 10 克
- 紅棗 6 粒
- 豬膶 250 克

調味

- 海鹽半茶匙

做法

1 蟲草花、雪耳、杞子分別浸洗，雪耳去蒂；紅棗去核；豬膶切塊，出水。
2 將雪耳、紅棗、豬膶放入煲內，用 1.5 公升水煮個半小時，加入蟲草花、杞子煮 10 分鐘，調味即可供服。

食療功效

補益肝腎、潤膚養顏。適合工作疲勞、經常捱夜、臉色晦暗或有色斑、肝腎陰虛者服用。

飲食宜忌

本品清香可口，不寒不燥，過肥過瘦者、免疫力低者均可服；感冒發燒者少服。

認識主料

蟲草花：金黃色，外觀最大的特點是沒有蟲體，只有橙色或者黃色的「草」。有滋肺補腎護肝、抗氧化、抗衰老等功效。蟲草花的價值，視乎草頭的大小，越飽滿越大者為佳；乾扁、幼細及沒有草頭的蟲草花，食療價值很低。

戶外工作人士的調養

體力勞動者每天需要消耗很多熱量，一些從事車輛、電梯、冷氣機等維修的工人，每日要在漆黑的環境中「捐窿捐罅」，弄致蓬頭垢面，油漬斑斑，如此辛苦工作，一到休息用膳時間，他們大多會大魚大肉，認為這樣吃法才可應付身體熱量的支出。

地盤工、外展工作者，工作環境酷熱，沙塵滾滾，體力消耗大，故碟頭飯，大杯凍飲等是他們最常用的午膳；放工後工人們尤其最愛吃燒味，肥美的燒鵝、燒腩肉可以吃下一整碟，大大枝啤酒下肚才有滿足感。正是這種暴飲暴食的飲食習慣，令他們體內血脂、膽固醇超標，不經不覺擁有個大肚腩，漸漸成為三高症的族群。

飲食不節，嗜酒無度，過量食辛熱厚味，會使痰熱內蘊。加上酷熱塵多的污濁環境，風熱邪毒，內外夾迫，身體就易生毛病。在充滿塵埃場所工作的人，患上矽肺病的比率亦會偏高，要預防這類肺部毛病，日常宜多飲一些補充能量、清肺熱、健脾祛濕的茶飲及湯水作保健之用。

魚腥草

又名蕺菜，鮮品有濃烈魚腥味，煮熟味清香。含揮發油，故不宜久煎。長期吸煙、痰黃黏稠者可常服。

桔梗

有鎮咳及增強抗炎作用，又能降血糖、降膽固醇。如買不到新鮮魚腥草，可用桔梗 3 克，配甘草 5 克泡茶服，對呼吸道亦能起到良好的保護作用。

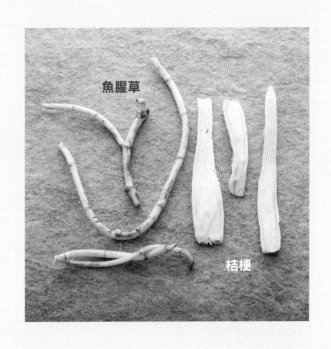
魚腥草

桔梗

魚腥草桔梗茶

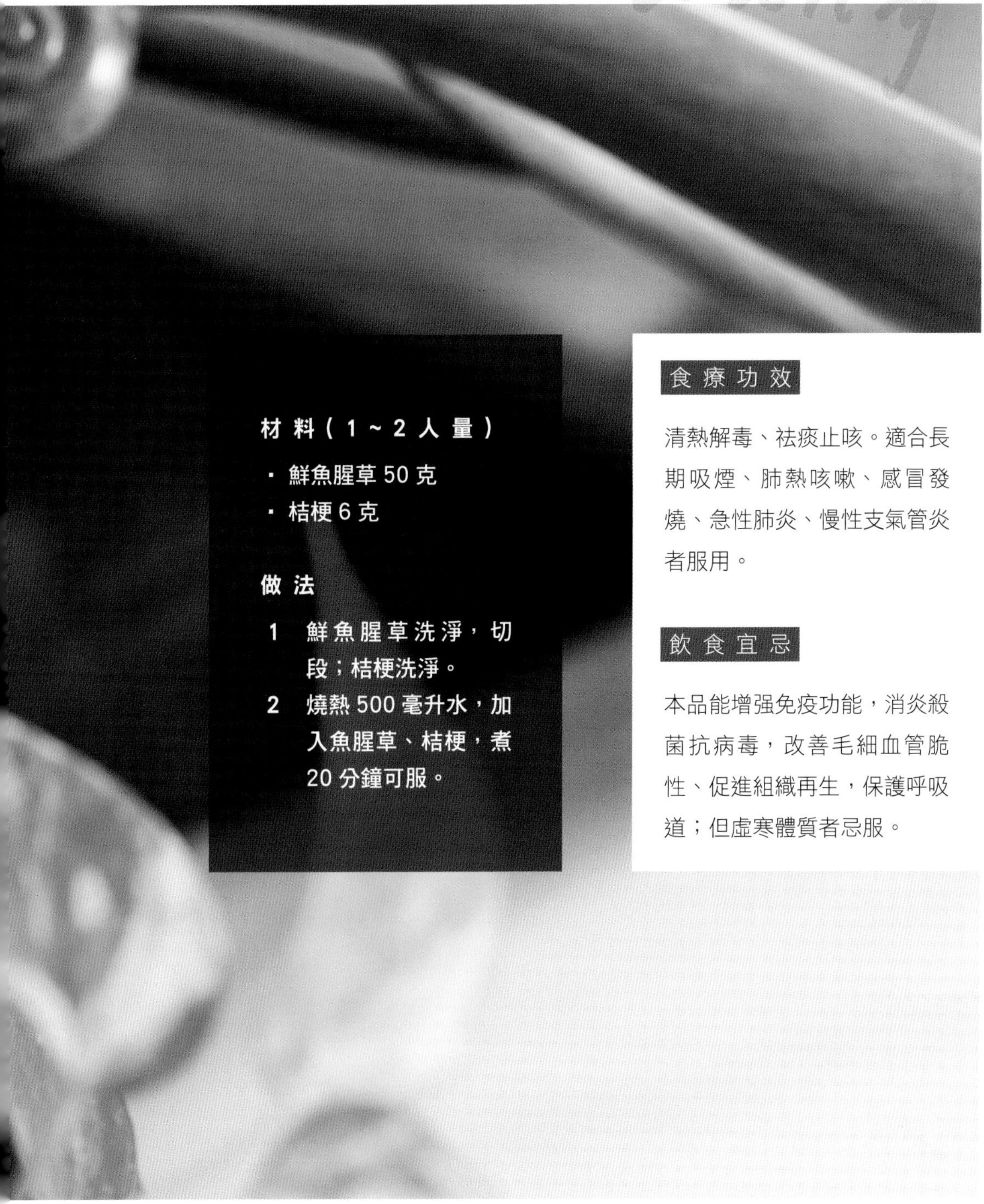

材料（1～2 人量）

· 鮮魚腥草 50 克
· 桔梗 6 克

做 法

1 鮮魚腥草洗淨，切
  段；桔梗洗淨。
2 燒熱 500 毫升水，加
  入魚腥草、桔梗，煮
  20 分鐘可服。

清熱解毒、祛痰止咳。適合長
期吸煙、肺熱咳嗽、感冒發
燒、急性肺炎、慢性支氣管炎
者服用。

飲 食 宜 忌

本品能增強免疫功能，消炎殺
菌抗病毒，改善毛細血管脆
性、促進組織再生，保護呼吸
道；但虛寒體質者忌服。

葛花茶

材料（1人量）

‧ 葛花 5 克

做法

1 葛花放入壺內，用開水先沖洗一遍，將水倒走。

2 再注入 250 毫升開水，焗5 分鐘可服用。

認識主料

葛花： 為豆科植物野葛的乾燥花蕾，主治傷酒發熱煩渴、不思飲食。在中藥店有售，以乾燥、體輕、無臭、味淡、無生蟲者為佳。

本品能解酒毒，對醉酒、飲酒過度及痰濕體質者很有幫助；但脾胃虛寒者慎服。

食 療 功 效

解酒醒脾、清熱除煩。適合飲酒過量、頭痛頭暈、心神煩亂、胸膈痞塞、嘔吐酸水者飲用。

馬齒莧瘦肉湯

材料（3～4人量）

- 新鮮馬齒莧 250 克
- 雞蛋 2 個
- 瘦肉 300 克

調味

- 海鹽半茶匙

做法

1. 馬齒莧洗淨，切段；雞蛋打散；瘦肉洗淨，切片後出水。
2. 瘦肉先用 800 毫升水煮至大滾，加入馬齒莧，滾 20 分鐘，最後加入蛋汁及調味，蛋花浮起即可供食。

食療功效

清熱解毒、散血消腫。適合在沙塵滾滾或日曬雨淋、工作環境欠佳的人士常服。

飲食宜忌

本品清香味微酸，常服能防止肺結節形成，避免患上矽肺病；但脾胃虛寒、腹瀉者及孕婦忌服。

認識主料

新鮮馬齒莧：能消炎殺菌，預防矽肺病。在山草藥檔經常有售，以新鮮、質嫩、整齊少碎、無雜質者為佳。

健脾益胃、化痰止咳。適合身
體發熱、濕火骨痛、關節疼痛
者服用。

本品滋補而不膩滯，不寒不
熱，老少皆宜。一般體質均可
服，但脾胃虛寒者宜少食。

野葛菜蜜棗
生魚湯

材料（3～4人量）

- 野葛菜 300 克
- 蜜棗 2 粒
- 生薑 2 塊
- 生魚 1～2 條

調味

- 海鹽半茶匙

做法

1. 野葛菜去根，洗淨，切段；蜜棗沖洗。
2. 生魚劏洗淨，用少許油煎香。
3. 燒熱 1.7 公升水，加入薑、生魚、蜜棗煮至大滾，再加入野葛菜，改用文火煮 2 小時，調味即成。

認識主料

野葛菜：有鎮咳祛痰作用，對支氣管炎、肺炎球菌、金黃葡萄球菌、大腸桿菌等均有抑制作用。對吸入灰塵多，易患氣管及肺臟疾病者最為有益。

生魚：因其生命力強，能在離水後生存相當一段時間，故稱「生魚」。其肉質堅實，味道鮮美，有利水祛痰、催乳補血的作用。最好選購頭大、身小、顏色較鮮明的野生品種生魚，療效較佳。

腐竹馬蹄髮菜蠔豉湯

材料（3～4人量）

- 腐竹 1 張
- 馬蹄 10 粒
- 髮菜 10 克
- 蠔豉 100 克
- 生薑 3 片
- 瘦肉 250 克

調味

- 海鹽半茶匙

做法

1 腐竹沖淨；馬蹄去皮、洗淨，切開；髮菜浸洗。

2 蠔豉洗淨，與切了片的瘦肉同出水。

3 材料放入煲內，用 1.7 公升水煮至大滾，改用文火煮 2 小時，調味即成。

髮菜

蠔豉

認 識 主 料

髮菜： 髮菜有良好的理腸去垢功效。以色澤烏黑、髮絲細長、乾淨無雜質者為佳。但市場有以植物根鬚染成黑色，混充髮菜，可用暖水浸泡片刻，看其是否發脹或脫色加以區別。

蠔豉： 蠔豉所含的多種氨基酸有解毒作用，可以除去體內有毒物質。蠔豉又具補血功效，含鋅量高，對貧血、孕婦、胎兒、更年期婦女均甚為有益。以生曬金蠔味最香，但由於蠔豉無論怎樣曬仍有一定濕度，故海味店放在凍櫃中儲存的生曬蠔豉質較優，因不需要用防腐劑。

食療功效

清熱化痰、軟堅散結。適合陰虛內熱、發熱心煩、甲狀腺腫大、淋巴結核、高血壓、便秘人士服用。

飲食宜忌

本品滋補美味，又有理腸除垢功效，在塵大、空氣差的環境工作人士可常服。一般體質者皆可服，但脾胃虛寒者及有皮膚病患者不宜多食。

久 坐 及 久 站 人 士 的 調 養

有些工種需要長時間在冷氣間站立或久坐,如教師、售貨員、超市的收銀員、酒樓接待員等;職業司機、銀行櫃枱、文書工作、售票員等則要坐着工作。無論久站或久坐,室溫太冷,時間太長,都容易誘發各種毛病。

腳部水腫、靜脈曲張與久坐、久站都有密切關係。坐得太久,長時間不運動會引致頸椎病、下肢靜脈栓塞、肥胖、便秘、腰肌勞損等多種病症;而站得太久,易引致椎間盤突出、坐骨神經痛、脊椎側彎、下肢靜脈栓塞等毛病。因此,因工作關係而需要長時間在冷氣間站立,或長時間在冷氣間坐着的打工男女,要避免肩頸痛、腰腿痛等毛病。可以隨身帶備小披肩或薄外套,在冷氣大的地方覆蓋膝蓋、頸椎、肩膊等關節以防邪氣入侵。冷氣房的濕度低,長時間逗留要多飲水,以防皮膚乾燥及眼睛乾澀。做運動和曬太陽可以增強抵抗力,排汗以驅逐體內寒濕之氣;一星期兩至三次用溫熱水或薑水泡腳也是減少患上「冷氣病」的方法。打工仔可根據自己的工作及生活環境,作出行為上的改善。工作環境不可能即時作改變,可多用食療來作調理。

認 識 主 料

香茅

無論鮮品、乾品均有濃烈檸檬香味，可用於湯類、肉類食品的調味料。但所提煉的香茅精油：六磷酸葡萄糖去氫酵素缺乏症（G6PD），俗稱蠶豆症人士忌用。

蘋果

能調節脂肪代謝，增強腸道功能。所含微量元素銅，對人體血液、中樞神經和免疫系統均有好處。選購以果形圓整、臍部寬大成熟、果皮亮麗、色澤均勻，用食指彈有清脆聲為佳品。

香茅

蘋果

香 茅 蘋 果 茶

材料（2人量）

- 香茅 3 支
- 蘋果 1 個
- 生薑 4 片

調味

- 紅糖 1 湯匙

做法

1 香茅洗淨，切片；蘋果去皮、去核、切粒。

2 將 700 毫升水煮滾，加入香茅、生薑、蘋果煮 20 分鐘，加入紅糖煮溶即成。

食療功效

疏風解表、祛瘀通絡。適合脘腹冷痛、筋骨疼痛、頭痛、消化不良、氣滯人士飲用。

飲食宜忌

本品清香味美、暖胃祛寒，常服能防膽固醇高及防膽結石；陰虛內熱者宜少服。

加味銀花茶

材料（2 人量）

· 金銀花 10 克
· 玄參 10 克
· 當歸 6 克
· 甘草 3 克

做法

1　將全部材料浸洗。
2　用 700 毫升水把全部材料
　　煮 20 分鐘即成。

認識主料

金銀花：見 P110「預防眼睛疲
勞的食療」篇──夏枯草金銀花
茶

當歸：見 P95「預防低血壓的
食療」篇──當歸黃蓍雞蛋茶。

玄參：玄參苦泄滑腸而通便，瀉
火解毒而利咽，臨床應用範圍較
為廣泛，但到底是藥，不宜作長
服的滋補劑。選購以質堅實、斷
面色黑、乾燥無蟲蛀者為佳。

食療功效

養陰清熱、活血通絡。適
合長期在冷氣間站立或坐
着工作，雙腳浮腫、腳重
重、容易抽筋、血流不
暢、腸燥便秘者服用。

飲食宜忌

本品味甘微苦，有助預防
靜脈栓塞；但脾胃虛寒、
泄瀉者及孕婦忌服。

當歸　玄參　金銀花

桂枝乾薑紅棗茶

材料（2 人量）

- 桂枝 6 克
- 乾薑 6 克
- 紅棗 6 粒

調味

- 蜂蜜 1 湯匙

做法

1 紅棗去核，切片，與桂枝、乾薑同沖洗。

2 將材料放入煲內，用 600 毫升水煮 15 分鐘，待溫加入蜂蜜調味即成。

食療功效

溫經散寒、養血通脈。適合長期在冷氣間工作、畏寒肢冷、筋骨酸痛、脾虛泄瀉及陽虛體質者服用。

飲食宜忌

本品能除寒通氣，逐風濕痹痛；但陰虛內熱、血熱妄行及孕婦忌服。

桂枝

乾薑

認識主料

桂枝：是肉桂樹的乾燥嫩枝，有發汗解肌、溫經通脈、散寒止痛等功效。但溫熱病、陰虛火旺及出血症忌用。以枝條嫩細均勻，色紅棕，香氣濃者為佳。

乾薑：是生薑採收後經處理乾淨，切片曬乾或低溫烘乾而成，性味辛熱燥烈，能溫中散寒、健運脾陽。以乾燥、氣香、無蟲蛀者為佳。

杜仲核桃
補腎湯

材料（3～4人量）

· 杜仲 30 克
· 杞子 6 克
· 核桃肉 50 克
· 熟地 20 克
· 生薑 3 片
· 豬腰 1 對

調味

· 海鹽半茶匙

做法

1　杜仲、杞子、核桃肉、熟地分別浸洗；豬腰去盡白筋膜，切片後漂洗，出水。

2　將杜仲、核桃肉、熟地、薑片及豬腰放入煲內，用 1.5 公升水煮 1 小時，加入杞子煮 5 分鐘，調味即成。

認 識 主 料

豬腰：豬腰清洗竅門：將豬腰子剝去薄膜，剖開，剔去筋，切成所需的片或花，用清水漂洗一遍，撈起瀝乾。豬腰用米酒拌和、捏擠，用水漂洗兩三遍，再用開水燙一遍，即可去羶臭味。

杜仲：能降血壓、補肝腎，強筋骨，安胎氣。以皮厚而大、外面黃棕色、內面黑褐色而光，折斷時白絲多者為佳。

食療功效

補益肝腎、強壯筋骨。適合陽虛、肝腎虧虛、耳鳴眩暈、腰酸背痛、雙腳軟弱無力者服用。

飲食宜忌

本品補肝腎、強筋骨，在冷氣間長期站立工作者可常服。一般體質皆可用，但痛風、高膽固醇人士不宜用豬腰，可用瘦肉 250 克代替。

雞血藤紅棗雞肉湯

材料（3～4 人量）

- 雞血藤 30 克
- 紅棗 8 粒
- 生薑 3 片
- 雞胸肉 250 克

調味

- 海鹽半茶匙

做法

1. 雞血藤浸洗；紅棗去核；雞胸肉洗淨，切塊後出水。
2. 將全部材料放入煲內，用 1.5 公升水煮至大滾，改用文火煮 1 小時，調味即成。

食療功效

活血舒筋、養血調經。適合氣虛血弱、面色萎黃、手足麻木、風濕痺痛、婦女痛經、閉經者服用。

飲食宜忌

本品強筋活絡、補血、暖腰膝，對長期在冷氣間工作者很適合；但陰虛火旺者慎服。

認識主料

雞血藤：當它的莖被切斷以後，其木質部就立即出現淡棕紅色，不久慢慢變成鮮紅色液流出，像雞血，故名。有散氣、活血、舒筋、活絡等功效。宜儲存在通風乾燥處，防霉、防蛀。

密 封 辦 公 室 工 作 人 士 的 調 養

寫字樓租金貴，寬敞開揚的辦公室並不太多，大部分打工仔都會在空氣不太流通的冷氣間工作，假如空調的保養、維修、清洗不足，對員工身體自然有不良影響。

密封辦公室會有不少廢氣和微塵粒，影印機所產生的微粒，電腦釋放出的輻射，實在不容忽視。很多打工一族經常喉痛、鼻敏感、傷風感冒，都可能由於辦公室空氣中的二氧化碳含量過高，缺乏新鮮空氣引致。而長期在這種環境下工作，也易令人精神緊張、肌肉疲勞痠痛、頭昏氣悶。

既然環境不易改變，不妨在辦公室擺放些較易種植的「天然空氣清新劑」，如虎尾蘭、仙人掌、蘆薈一類小盆栽。細細盆的矮腳虎尾蘭，放氧量較其他植物高出30倍，起到清新空氣的作用；迷你蘆薈小盆栽雖然袖珍，卻是抗輻射能手，且能阻擊甲醛、分解影印機等排放出來的苯；仙人掌球耐旱易於打理，是吃微生物及吸附灰塵的高手，可起到淨化環境的作用。

當然，黃金葛、吊蘭、鐵樹、富貴竹、龜背竹等都同樣能淨化空氣，具有吸油煙、清除空氣中的微生物、吸微塵等功效，但這些植物一般都需要較多陽光、水分及時間打理，未必適合放置在辦公室。故除了用天然植物淨化環境，日常多用食療調養對健康最為有益。

認識主料

黑木耳

新鮮黑木耳含有一種卟啉的光感物質，食用後經太陽照射可引起皮膚瘙癢、水腫；乾品木耳則食用安全。選購以耳片烏黑光潤，背面呈灰白色，片大均勻，浸發後耳瓣舒展，體輕質潤，半透明，脹性好，無雜質，有清香氣味者為佳品。

黑木耳薑棗茶

材料（2 人量）

- 黑木耳 20 克
- 生薑 30 克
- 紅棗 6 粒

調味

- 紅糖 30 克

做法

1 黑木耳用清水浸軟，去蒂後切小塊；生薑切絲、紅棗去核。

2 將材料放入煲內，用 700 毫升水煮 30 分鐘，加入紅糖煮溶即成。

食療功效

暖胃祛寒、淨化血液。適合心腦血管疾病、貧血、血脈不通、寒濕性腰腿疼痛、便秘者服用。

飲食宜忌

本品膠質豐富，含鐵、鈣等營養素很高，是女性及一般人士預防貧血及骨質疏鬆佳品；但有出血性疾病、腹瀉者忌服，孕婦不宜多服。

珊 瑚 草 杞 子 茶

材料（2 人量）

· 珊瑚草 15 克
· 杞子 5 克

調味

· 冰糖 50 克

做法

1 珊瑚草浸至完全透明，洗淨後切段；杞子浸洗。
2 將珊瑚草用 600 毫升水煮 20 分鐘，加入杞子再煮 5 分鐘，下冰糖煮溶即成。

食療功效

消炎抗菌、淨化血液。適合皮膚粗糙、過敏、臉色黯淡憔悴、消化不良、口臭、便秘及容易上火者服食。

飲食宜忌

本品常服能增強血液循環，清宿便，排走體內毒素。但珊瑚草性寒涼，陽虛體質者不宜多食。

認 識 主 料

珊瑚草： 有「海燕窩」之稱，含大量鈣、鐵、磷、鎂、鉀、硒、錳等礦物質，是極佳的海洋營養食品。選購以天然曬乾的淺啡色、乾身無泥砂者為佳。

杞子： 見P31「氣虛體質」篇──黃蓍杞子紅棗茶。

杞子
珊瑚草

絞股藍蜂蜜茶

材料（2人量）

· 絞股藍 10 克

調味

· 蜂蜜 1 湯匙

做法

1 將絞股藍放入壺內，用開水沖洗一遍，將水倒走。

2 再注入 500 毫升開水，焗 10 分鐘後加入蜂蜜即成。

食療功效

消炎解毒、延緩衰老。適合體虛乏力、高血壓、高血脂、病毒性肝炎、慢性腸胃炎、慢性氣管炎者服用。

飲食宜忌

本品有清香草青味，對身體有固本培元功效，同時可防癌抗癌。但少數人服藥後會出現噁心嘔吐或頭暈眼花症狀，脾胃虛寒者忌服。

認識主料

絞股藍：優質絞股藍沖泡後捲縮的嫩葉、嫩莖會迅速張開，複葉五枚，茶水色澤嫩綠，清澈透明，多次沖泡依然不改其色，味略清苦卻有甘醇回味。差者莖老葉粗，甚者藤葉長連，茶色淡黃。

土茯苓赤小豆薏米龜板湯

材料（3～4 人量）

- 鮮土茯苓 60 克
- 赤小豆 30 克
- 薏米 30 克
- 蜜棗 2 粒
- 龜板 200 克
- 瘦肉 250 克

調味

- 海鹽半茶匙

做法

1. 鮮土茯苓切片後浸洗淨；赤小豆、薏米浸洗。
2. 龜板浸洗，出水；瘦肉切片，洗淨出水。
3. 將全部材料放入煲內，用 1.7 公升水煮至大滾，改用文火煮 2 小時，調味即成。

土茯苓：可除濕解毒、又可通利關節，及去汞毒（即水銀）。一般曬乾切片使用，街市有新鮮的土茯苓售賣，若切片生用，化濕解毒的功效會更強。現時很多山草藥檔檔主肯代客去皮及切片。

食 療 功 效

清熱排毒、滋陰補虛。適合陰虛內熱、腎虛腰膝酸軟、風濕疼痛、骨蒸勞熱、皮膚過敏者服用。

飲 食 宜 忌

本品為清熱解毒常用之保健湯水，適合在空氣質素欠佳的環境工作人士常服；但陽虛、脾胃虛寒及孕婦忌服。

土茯苓　　　　　龜板

龜板：能滋腎潛陽、益腎健骨、養血補心。以質乾燥、板上有血斑、塊大無腐肉者為佳。

猴頭菇冬菇雞湯

食療功效

固本扶正、強壯體質。適合氣虛消
化不良、胃脹胃痛、神經衰弱、臉
色晦暗、二便不暢者服用。

飲食宜忌

本品有助排毒養顏，強壯體質，促進
人體新陳代謝，加強體內廢物排泄；但
脾胃寒濕、氣滯或皮膚瘙癢人士忌食。

認 識 主 料

猴頭菇： 對消化系統毛病有
良好療效，亦有助治療輕度
神經衰弱。選購以色澤淺黃、
質嫩肉厚、完整無缺、乾燥
無蟲蛀者為佳。

材 料（3 ～ 4 人 量）

- 猴頭菇 3 朵
- 冬菇 5 ～ 6 朵
- 海藻 5 克
- 生薑 3 片
- 土雞 1 隻

調 味

- 海鹽半茶匙

做 法

1　猴頭菇、冬菇、海藻分別
　　浸洗，猴頭菇、冬菇去蒂。
2　土雞劏洗淨，斬大件後與
　　猴頭菇齊出水。
3　將全部材料放入煲內，用
　　1.7 公升水煮至大滾，改用
　　文火煮 2 小時，調味即成。

海藻： 具有清熱、軟堅散結
等功效。所含食物纖維有助
消化及促進廢物排泄，減少
毒素積聚體內；以螺旋海藻
營養較佳。

CONTENTS

Ginger and red date tea with
dark brown sugar (2 servings) Ref. p.013

This tea also helps ease menstrual discomfort and regulate menstrual cycles. Consuming this tea in spring and autumn helps prevent common cold and flu; in summer helps cures vomiting and diarrhoea; in winter helps warm the body and expels Coldness. However, consumption through prolonged period may lead to accumulated Htateat in the body, depleting the Yin, and undermining eyesight. Those with Yin-Asthenia accompanied by accumulated Heat, and haemorrhoid patients should avoid.

Ingredients

- 15 g sliced mature ginger
- 15 g red dates (de-seeded and sliced)

Seasoning

- 2 tsp dark brown sugar (Kurozatou)

Method

1 Put all ingredients into a teapot. Pour in boiling water. Swirl to rinse well. Drain.
2 Put dark brown sugar into the same teapot. Pour in 500 ml of boiling water. Cover the lid and leave it for 10 minutes. Strain and serve.

Mutton soup with fresh yam and
dried longans (3 to 4 servings) Ref. p.016

The dried longans in this soup help regulate blood chemistry and calm the nerves. Those with Qi- and Blood-Asthenia may feel free to consume. Those who tend to build up Fire in the body easily may use 2 to 3 water chestnuts instead of dried longans. Those who haven't fully recovered from common cold or flu and those with Yin-Asthenia accompanied by accumulated Heat should not consume.

Ingredients

- 150 g fresh yam
- 500 g fresh mutton

- 30 g ginger
- 12 g dried longans (shelled and de-seeded)

Seasoning

- 1/4 tsp ground white pepper
- 1/2 tsp sea salt

Method

1 Chop mutton into chunks. Blanch in boiling water. Drain and rinse well. Drain again.
2 Peel and rinse fresh yam. Cut into chunks. Soak and rinse dried longans in water. Slice the ginger.
3 Put all ingredients into a pot. Add 1.7 litre of water. Bring to a vigorous boil over high heat. Turn to low heat and simmer for 2 hours. Season with salt and pepper. Serve both the soup and the solid ingredients.

Luo Han Guo and Mai Dong tea
(2 servings) Ref. p.019

This tea is great for those who need to talk a lot on their job, such as teachers, vocalists and salespersons as it helps prevent chronic pharyngitis. However, those of Asthenia-Coldness body type should not consume.

Ingredients

- 6 g Luo Han Guo
- 10 g Mai Dong

Method

1 Put all ingredients into a teapot. Pour in boiling water and swirl to rinse well. Drain.
2 Pour in 500 ml of boiling water. Cover the lid and leave it for 15 minutes. Strain and serve.

Teal soup with Sha Shen and Yu Zhu

(3 to 4 servings) Ref. p.022

This soup is great for those who have to stay up late at night and those having Asthenia whose body fails to pick up the potent components in health tonics. However, those with coughs due to Wind-Coldness and diarrhoea due to Spleen-Asthenia should use with care. Those with coughs due to Phlegm-Heat should avoid.

Ingredients

- 15 g Sha Shen
- 12 g Yu Zhu
- 5 g white fungus
- 5 g Goji berries
- 1 piece dried tangerine peel
- 1 teal

Seasoning

- 1/2 tsp sea salt

Method

1 Dress the teal and rinse well. Chop into pieces and blanch in boiling water. Drain.
2 Soak and rinse Sha Shen, Yu Zhu, dried tangerine peel, Goji berries and white fungus in water. Cut off the tough root of the white fungus.
3 Put all ingredients (except Goji berries) into a soup pot. Add 1.5 litre of water. Bring to a vigorous boil over high heat. Turn to low heat and simmer for 30 minutes. Add Goji berries and boil for 5 more *minutes. Season with salt. Serve both the soup and solid ingredients.

Yun Ling and Bai Zhu tea (2 servings)

Ref. p.025

This tea is especially good for those who want to consume health tonics, but haven't fully recovered their Spleen-Stomach functions after influenza. This tea would help strengthen their Spleen functions and expel Dampness. The nutrients in health tonics can be absorbed a lot more easily. However, those with Yin-Asthenia and without any Dampness-Heat accumulated should not consume.

Ingredients

- 7.5 g Yun Ling
- 7.5 g Bai Zhu
- 7.5 g Bian Dou Yi (hyacinth bean coat)
- 7.5 g liquorice

Method

Soak and rinse the ingredients. Cook in 700 ml of boiling water for 30 minutes. Strain and serve.

Pork rib soup with winter melon, small red beans and Job's tears (3 to 4 servings) Ref. p.028

This soup helps alleviate gout and muscle cramps. However, those with Asthenia-Coldness in the Spleen and Stomach and those with frequent urinations at night should consume in moderation.

Ingredients

- 500 g winter melon
- 50 g small red beans
- 50 g Job's tears
- 2 candied dates
- 300 g pork ribs

Seasoning

- 1/2 tsp sea salt

Method

1 Chop pork ribs into pieces. Rinse and blanch in boiling water. Drain.
2 Soak and rinse small red beans and Job's tears in water. Rinse winter melon with its skin on. Chop into pieces.
3 Put all ingredients into a soup pot. Add 1.7 litres of water. Bring to a vigorous boil over high heat. Turn to low heat and simmer for 2 hours. Season with salt. Serve.

Huang Qi, Goji and red date tea

(2 servings) Ref. p.031

This tea helps strengthen the lungs and boost immune system in the flu season. It is suitable for all ages. However, those with overwhelming exogenous Evil and accumulated Heat-toxicity syndrome should not consume.

Ingredients

- 15 g Huang Qi (a.k.a. Bei Qi)
- 6 g Goji berries
- 6 red dates

Method

1 Finely chop the Huang Qi. Rinse Huang Qi and Goji berries in water. De-seed the red dates. Slice them.
2 Put all ingredients into a tea pot. Pour in boiling water to cover. Drain to rinse them. Add 500 ml of hot boiling water. Cover the lid and leave them for 15 minutes. Serve.

Double-steamed fish maw soup with Dang Shen and Goji berries (2 to 3 servings) Ref. p.034

This soup is good for both sexes, especially those who had surgical operations in recent past. It helps enhance skin texture and make your cheeks rosy. However, those with poor Qi circulation in the Spleen and Stomach meridians, those with overwhelming Fire in the Liver, and those not fully recovering from flu due to exogenous pathogenic factor should not consume.

Ingredients

- 15 g Dang Shen
- 6 g Goji berries
- 3 slices ginger
- 6 red dates
- 250 g lean pork
- 120 g rehydrated fish maw

Seasoning

- 1/2 tsp sea salt

Method

1 Rinse and slice lean pork. Blanch lean pork and fish maw in boiling water. Drain and set aside. De-seed the red dates. Soak and rinse Dang Shen and Goji berries.
2 Put all ingredients into a double-steaming pot. Pour in 600 ml of boiling water. Double-steam for 2 hours. Season with salt. Serve.

Honey haw tea (2 servings) Ref. p.037

This tea eases indigestion and aids lipid digestion. However, haw is too sour by itself and may damage the lining of stomach. On the other hand, honey nourishes Yin, moistens Dryness, and promotes Qi (vital energy) flow in the Spleen meridian. It eases the irritation on stomach lining caused by haw. The elderly may add black tea or Pu Er tea leaves to this tea to help digestion. However, pregnant women, those with low blood sugar level, Spleen- or Stomach-Asthenia, or stomach ulcer should not consume.

Ingredients

- 30 g haw

Seasoning

- 1 tbsp honey

Method

1 Put haw into a teapot. Pour in boiling water. Swirl to rinse the haw. Drain.
2 Add 500 ml of boiling water. Cover the lid and let it sit for 10 minutes. Stir in honey. Serve.

Chicken soup with Yi Mu Cao and black beans (2 to 3 servings) Ref. p.040

This soup eases blood stasis and promotes new tissue generation. It also has antioxidant, anti-ageing, and skin-beautifying effects. However, women should not consume during menstruation and pregnant women should not consume at all.

Ingredients

- 30 g Yi Mu Cao (Chinese motherwort)
- 50 g Green-kernel black beans
- 6 red dates
- 2 slices ginger
- 150 g chicken breast

Seasoning

- 1/2 tsp sea salt

Method

1 Slice the chicken breast. Blanch in boiling water. Drain and set aside. Rinse the Yi Mu Cao.
2 Soak black beans in water until soft. De-seed the red dates.
3 Put all ingredients into a clay pot. Add 800 ml of water. Bring to the boil over high heat. Then turn to low heat and cook for 1 hour. Season with salt. Serve.

Albizzia bergamot tea (2 servings) Ref. p.043

This tea alleviates restlessness, depression and insomnia. Frequent consumption helps ease nervousness, calms the nerves and reduces mental fatigue. It also helps cure acute conjunctivitis or blurry vision. However, pregnant women should not consume.

Ingredients

- 5 g He Huan Hua (dried albizzia flowers)
- 10 g dried bergamot

Seasoning

- 1 tbsp honey

Method

1 Put both He Huan Hua and dried bergamot into a teapot. Pour in boiling water and swirl to rinse the ingredients. Drain.
2 Pour in 500 ml of boiling hot water. Cover the lid and leave them for 10 minutes. Stir in honey. Serve.

Sliced pork soup with day lily flowers and cloud ear fungus

(3 to 4 servings) Ref. p.046

This soup is delicious, aromatic and nutritious. It is suitable for all ages and it clears Heat in the Lungs while regulating Qi (vital energy) flow in the Liver. It is suitable for all people with different predisposed body types.

Ingredients

- 15 g day lily flowers
- 6 g cloud ear fungus
- 2 eggs
- 250 g lean pork
- 2 sprigs coriander

Seasoning

- 1/2 tsp sea salt

Method

1 Rinse day lily flowers and cloud ear fungus. Make a knot on each day lily flower. Cut off the tough base of cloud ear fungus. Slice the pork. Blanch pork, day lily flowers and cloud ear fungus in boiling water together. Drain.
2 Whisk the eggs. Rinse coriander. Finely chop it.
3 Boil 700 ml of water in a pot. Put in day lily flowers, cloud ear fungus and sliced pork. Boil for 10 minutes. Stir in whisked egg. Sprinkle with chopped coriander. Season with salt. Bring to the boil and serve.

Lotus leaf and bitter melon tea

(2 servings) Ref. Ref. p.049

This tea is mildly bitter in taste, suitable for those with diabetes, Dampness-Heat, obesity, oedema and those working under high temperature. However, those with Asthenia-Coldness body type and those with skinny build and Asthenia should use with care.

Ingredients

- 10 g dried lotus leaf
- 20 g dried bitter melon

Method

1 Put dried lotus leaf and bitter melon into a teapot. Pour in boiling water. Swirl to rinse ingredients. Drain.
2 Pour in 500 ml of boiling water. Cover the lid and leave it for 10 minutes. Serve.

Egg drop soup with laver and tofu

(3 to 4 servings) Ref. p.052

This is a nutritious soup that doubles as a therapeutic food for those with Dampness-Heat and obesity. However, those with gout, Spleen- or Stomach-Asthenia should not consume.

Ingredients

- 1 small piece dried laver
- 2 cubes tofu
- 2 eggs
- 1 tbsp finely shredded ginger
- 1 tbsp finely chopped spring onion

Seasoning

- 800 ml stock
- 1/2 tsp sea salt

Method

1 Rinse the tofu. Coarsely shred it. Rinse the laver. Whisk the eggs.
2 Boil the stock in a pot. Put in tofu and shredded ginger. Bring to the boil. Add laver and stir in whisked eggs. Season with sea salt and sprinkle with spring onion at last. Serve.

Bitter orange blossoms tea with kumquat honey (2 servings) Ref. p.055

This tea is fragrant and zesty. It helps promote blood circulation and beautifies the skin, so that it is also called beauty tea. People of any body type may consume. But pregnant women should avoid.

Ingredients

- 6 g bitter orange blossoms

Seasoning

- 1 tbsp kumquat honey

Method

1 Put bitter orange blossoms into a teapot. Pour in boiling water. Swirl to rinse well. Drain.
2 Pour in 500 ml of boiling water. Cover the lid and leave it for 5 minutes. Stir in kumquat honey. Serve.

Jasmine tea with cloves (2 servings)

Ref. p.058

This tea strengthens the Stomach, eases stomach gas and bloating. It also freshens the breath and alleviates depression. However, those with Yin-Asthenia, accumulated Heat, bitterness and dryness in the mouth, and those with dry hard stool should not consume.

Ingredients

- 3 g cloves
- 5 g dried jasmine flowers
- 3 g green tea

Method

1 Put all ingredients into a teapot. Pour in boiling water and swirl to rinse the ingredients. Drain.
2 Pour in 500 ml of boiling water. Cover the lid and leave it for 5 minutes. Serve.

Cinnamon malt tea (2 servings) Ref. p.061

This tea is fragrant and sweet. Diabetics may also consume without worries as long as you omit the sugar. This tea helps strengthen the stomach and reduce blood glucose level. Yet, those with Yin-Asthenia and overwhelming Fire, or Sthenic-Fire should not consume. Pregnant women or lactating mothers should also avoid.

Ingredients

- 1/2 tsp ground cinnamon
- 30 g fried malt

Seasoning

- 2 tsp red sugar

Method

1 Rinse the fried malt. Drain.
2 Boil malt in 600 ml of water for 15 minutes. Add ground cinnamon and red sugar. Cook until sugar dissolves. Serve.

Haw tea with Jue Ming Zi (2 servings)
Ref. p.062

This tea helps weight loss and reduces blood fat level. It also clears the accumulated fata in the blood. It is suitable for all people. However, those with loose or water stool, peptic ulcer, excessive stomach acid and pregnant women should avoid.

Ingredients

- 15 g haw
- 20 g fried Jue Ming Zi

Seasoning

- 2 tsp red sugar

Method

1 Put all ingredients into a teapot. Pour in boiling water and swirl to rinse well. Drain.
2 Pour in 500 ml of boiling water. Add red sugar. Cover the lid and leave it for 15 minutes. Strain and serve.

Pork rib soup with dried mussels, Chinese marrow and Job's tears
(2 servings) Ref. p.064

This soup is flavourful and delicious. It benefits the Spleen and Kidneys while alleviating difficulty in urination or defecation. People of any body type may consume. Yet, pregnant women should avoid Job's tears and they should replace them with 30 g of black-eyed beans.

Ingredients

- 60 g dried mussels
- 2 Chinese marrows
- 30 g raw Job's tears
- 250 g pork ribs
- 3 slices ginger

Seasoning

- 1/2 tsp sea salt

Method

1 Soak and rinse the dried mussels in water. Drain. Scrape off the skin of the Chinese marrows with a knife. Cut into chunks. Soak and rinse the Job's tears. Blanch the pork ribs in boiling water. Drain.
2 Boil 1.5 litre of water. Put in all ingredients. Boil for 2 hours. Season with salt. Serve.

Fish tail soup with chayote and black-eyed beans (3 to 4 servings) Ref. p.066

This soup is sweet and delicious. It is suitable for all ages and people in different physical conditions. Those gout patients should not consume black-eyed beans. You can use 30 g of Job's tears instead.

Ingredients

- 2 chayotes
- 50 g black-eyed beans
- 1 piece dried tangerine peel
- 1 grass carp tail

Seasoning

- 1/2 tsp sea salt

Method

1 Peel the chayotes and cut into chunks. Soak and rinse the black-eyed beans and dried tangerine peel.
2 Rinse the fish tail. Fry in a little oil until golden on both sides.
3 Boil 1.5 litre of water in a pot. Put in all ingredients. Boil for 1 hour. Season with salt. Serve.

Lean pork soup with burdock, white fungus and sweet corn

(3 to 4 servings) Ref. p.068

This soup is sweet and tasty. It eliminates metabolic waste from the body, detoxifies and prevents cancer and stroke. However, those with Asthenia-Coldness in the Spleen and Stomach should consume in moderation.

Ingredients

- 150 g fresh burdock
- 10 g white fungus
- 2 ears sweet corn
- 2 slices ginger
- 250 g lean pork

Seasoning

- 1/2 tsp sea salt

Method

1 Scrub the burdock with skin on and rinse well. Cut into short lengths. Peel the sweet corn and rinse well. Cut into chunks. Soak white fungus in water until soft. Cut off the tough root.
2 Slice the pork and blanch in boiling water. Drain. Put all ingredients into a pot. Add 1.5 litre of water. Boil for 1 hour. Season with salt. Serve both the soup and the solid ingredients.

Clam soup with asparagus and bamboo fungus (3 to 4 servings) Ref. p.070

This soup has great seafood taste and helps ease restlessness, dissipates Heat, and improves eyesight. However, those with Asthenia-Coldness in the Spleen and Stomach, those not fully recovered from common cold and flu, and those having diarrhoea due to Yang-Asthenia should not consume.

Ingredients

- 150 g asparagus
- 20 g bamboo fungus
- 3 g Goji berries
- 300 g live clams (in shells)
- 1 tbsp shredded ginger

Seasoning

- 1/2 tsp sea salt

Method

1 Peel the asparaguses. Rinse and cut into short lengths. Soak and rinse the Goji berries. Soak bamboo fungus in water until soft. Cut off both ends. Blanch in boiling water and cut into short lengths. Blanch clams in boiling water until they open. Shell them and use the clam meat only.
2 Boil 1.2 litre of water in a pot. Put in all ingredients. Boil for 20 minutes. Season with salt and serve the soup with the solid ingredients.

Buckwheat tea with golden tickseed

(2 servings) Ref. p.073

This tea is rich in flavour with a clear and dark crimson colour. Those overweight who suffer from high blood pressure, blood glucose and blood fat levels may consume regularly. However, those with low blood pressure, Asthenia body type and allergy should use with care.

Ingredients

- 10 g dried golden tickseed
- 30 g fried buckwheat grains

Method

1 Put all ingredients into a teapot. Pour in boiling water and swirl to rinse them. Drain.
2 Pour in 500 ml of boiling water. Cover the lid and leave it for 15 minutes. Serve.

Dried guava green tea (2 servings) Ref. p.076

This tea is mildly bitter with a sweet aftertaste. The elderly and those in their middle age would benefit from it most. However, those with Asthenia-Coldness in the Spleen and Stomach, dry stool and constipation, diarrhoea and feeling full without eating.

Ingredients

- 30 g dried guava
- 5 g green tea loose leaves

Method

1 Put dried guava into a teapot. Pour in boiling water and swirl to rinse well. Drain.
2 Pour in 500 ml of boiling water. Add green tea leaves. Cover the lid and leave it for 7 minutes. Strain and serve.

Roselle apple tea (2 servings) Ref. p.078

This tea helps weight loss, slims the body and prevent high blood pressure, high blood glucose level and high blood fat level. Diabetics may use maple syrup instead of honey as maple syrup has lower glycemic index so that it does not raise blood glucose level drastically. However, those with excessive stomach acid should not consume. Women should also avoid during pregnancy and menstruation.

Ingredients

- 6 dried roselle flowers
- 1 apple

Seasoning

- 1 tbsp honey

Method

1 Rinse and peel the apple. Dice coarsely.
2 Boil the roselle flowers and diced apple in 600 ml of water for 15 minutes. Stir in honey before serving.

Tian Qi Hua tea (2 servings) Ref. p.080

This tea is sweet in taste with a hint of bitterness and it tastes a bit like ginseng. It helps prevents high blood pressure, blood glucose and blood fat levels. However, women should avoid during menstruation and pregnancy.

Ingredients

- 8 to 10 Tian Qi Hua

Method

1 Put Tian Qi Hua in a teapot. Pour in boiling water and swirl to rinse well. Drain.
2 Pour in 500 ml of boiling water. Cover the lid and leave it for 7 minutes. Strain and serve.

Fried black bean tea with
Du Zhong leaves (2 servings) Ref. p.082

This tea is sweet and aromatic. It tranquilizes the mind and relaxes. It is a great drink for those with Kidney-Asthenia and even pregnant women may consume it. However, the black beans impart more Heat after fried and those who suffer from overwhelming Fire easily, those with Yin-Asthenia and accumulated Fire and gout patients should avoid.

Ingredients

- 10 g Du Zhong leaves
- 50 g fried black beans with green kernels

Method

1 Put all ingredients into a teapot. Pour in boiling water and swirl to rinse well. Drain.
2 Pour in 500 ml of boiling water. Cover the lid and leave it for 15 minutes. Strain and serve.

Five-veggie soup for high blood
pressure (3 to 4 servings) Ref. p.084

This soup smells great and tastes delicious. It defies ageing, regulates pH level of body tissues, prevents and fights against cancer. However, those with Asthenia-Coldness in the Stomach and Spleen, and those with Yang-Asthenia should use with care.

Ingredients

- 60 g Chinese celery
- 200 g tomatoes
- 3 g spirulina
- 8 water chestnuts
- 1 onion

Seasoning

- 1/2 tsp sea salt

Method

1 Rinse Chinese celery and coarsely dice it. Rinse, peel and coarsely dice onion, water chestnuts and tomatoes. Rinse the spirulina.
2 Boil 800 ml of water. Put in all ingredients. Boil for 20 minutes. Season with salt. Serve.

Tofu thick soup with shiitake mushrooms and dried scallops

(3 to 4 servings) Ref. p.086

This soup if flavourful, delicious and nutritious. It is suitable for all ages. Only gout patients should avoid.

Ingredients

- 30 g dried shiitake mushrooms
- 3 dried scallops
- 1 cube tofu
- 1 tbsp frozen green peas
- 800 ml stock

Seasoning

- 1/2 tsp sea salt
- 1/2 tsp ground white pepper

Thickening glaze (mixed well)

- 1 tbsp caltrop starch
- 1 tbsp water

Method

1. Soak mushrooms and dried scallops in water until soft. Finely shred the mushrooms. Tear the dried scallops into shreds.
2. Rinse and finely shred the tofu. Thaw the green peas. Blanch in boiling water briefly. Drain.
3. Boil the stock in a pot. Put in shredded shiitake mushrooms and dried scallops. Cook for 20 minutes. Add tofu and green peas. Season with salt and pepper. Bring to the boil and stir in the thickening glaze. Cook while stirring continuously until it thickens. Serve.

Frog soup with pumpkin and garlic cloves (3 to 4 servings) Ref. p.088

This soup is sweet, delicious and nutritious. It protects the stomach lining while helping the excretion of toxins and heavy metal out of the body. It is suitable for all ages. Yet, those with Dampness accumulated in the body should not drink much.

Ingredients

- 250 g pumpkin
- 2 frogs (about 300 g)
- 50 g whole garlic cloves
- 1 tbsp shredded ginger

Seasoning

- 1/2 tsp sea salt

Method

1. Rinse the pumpkin with skin on. Cut into chunks. Peel the garlic cloves. Dress and rinse the frogs well. Chop into pieces. Blanch in boiling water. Drain.
2. Put all ingredients into a pot. Add 1.5 litre of water. Bring to a vigorous boil. Turn to low heat and simmer for 1 hour. Season with salt. Serve.

Pork pancreas soup with fresh yam, Goji berries and corn silks (3 to 4 servings)

Ref. p.090

This soup helps reduce blood glucose level and diabetics may consume regularly. Though it is good for people of any body type, those having fever due to common cold or flu should use with care.

Ingredients

- 200 g fresh yam
- 5 g Goji berries
- 30 g fresh corn silks
- 1 pork pancreas
- 2 slices ginger

Seasoning

- 1/2 tsp sea salt

Method

1 Peel and rinse the yam. Slice it. Soak and rinse the Goji berries and corn silks in water.
2 Trim off the fat on the pork pancreas. Rinse and slice it.
3 Boil 1.5 litre of water. Put in ginger, fresh yam and corn silks. Cook for 45 minutes. Put in pork pancreas and Goji berries. Boil for 15 minutes. Season with salt. Serve the soup and the solid ingredients.

Pork rib soup with kudzu, black-eyed beans and Job's tears

(3 to 4 servings) Ref. p.092

This soup is delicious and sweet. Those with high blood pressure, blood glucose and blood fat levels and those who consume alcohol excessively may consume regularly. However, those with Asthenia-Coldness in the Spleen and Stomach should use with care.

Ingredients

- 300 g kudzu
- 50 g black-eyed beans
- 30 g Job's tears
- 6 red dates
- 1 piece dried tangerine peel
- 300 g pork ribs

Seasoning

- 1/2 tsp sea salt

Method

1 Peel the kudzu. Rinse and cut into chunks. Soak and rinse black-eyed beans, Job's tears and dried tangerine peel in water. De-seed the red dates.
2 Rinse the pork ribs. Blanch in boiling water. Drain.
3 Boil 1.8 litre of water in a pot. Put in all ingredients. Bring to the boil over high heat. Turn to low heat and cook for 2 hours. Season with salt. Serve.

Dang Gui and Huang Qi tea with hard-boiled eggs (3 to 4 servings) Ref. p.095

This tea replenishes Qi and Blood. Women and those with Blood-Asthenia may consume regularly. However, those with Yin-Asthenia and overwhelming Fire, fever due to common cold or flu, Phlegm-Dampness or haemorrhagic diseases should not consume.

Ingredients

- 12 g head of Dang Gui root (sliced)
- 20 g Huang Qi
- 5 g Goji berries
- 2 eggs

Seasoning

- 1 tbsp red sugar

Method

1 Soak and rinse Dang Gui, Huang Qi and Goji berries in water. Put eggs with shells on in a pot of cold water. Cook over high heat until hard-boiled (for about 7 minutes after it starts to boil vigorously). Shell them.
2 Boil Dang Gui and Huang Qi in 700 ml of water for 30 minutes. Put in the shelled eggs, Goji berries and red sugar. Cook until sugar dissolves. Serve.

Dang Shen and Huang Jing tea

(2 servings) Ref. p.098

This tea is sweet in taste. Those who have depleted body fluid and Qi (vital energy) or Qi- and Blood-Asthenia may benefit from it most. However, those with diarrhoea due to Spleen-Asthenia, and those with poor Qi flow due to Phlegm-Dampness should not consume.

Ingredients

- 20 g Dang Shen
- 20 g Huang Jing
- 6 g fried liquorice
- 8 red dates

Method

1 Soak and rinse all ingredients. Slice Dang Shen thinly. De-seed the red dates.
2 Put all ingredients into a pot. Add 800 ml of water and boil for 30 minutes. Serve.

Ginseng tea with dried longans and red dates (2 servings) Ref. Ref. p.100

This tea is sweet and delicious. It promotes secretion of body fluid while boosting Qi (vital energy). It is suitable for both sexes. However, those with common cold or flu accompanied by fever, or constipation due to sthenic-Heat should not consume. When you consume this tea (or any food with ginseng for that matter), you should avoid having radish, strong tea or coffee on the same day.

Ingredients

- 10 g red ginseng rootlets
- 10 dried longans
- 6 red dates

Method

1 De-seed the red dates. Slice them. Put all ingredients into a teapot. Pour in boiling water and swirl to rinse the ingredients. Drain.
2 Pour in 500 ml of boiling water. Cover the lid and let them steep for 15 minutes. Strain and serve.

Beef shin soup with ginseng, Huang Qi and Wu Wei Zi (3 to 4 servings) Ref. p.102

This soup is delicious and flavourful. Those with anaemia, low blood pressure and those recovering after a recent surgical operation will benefit from it most. However, those having common cold or flu alongside fever, those with Yin-Asthenia and accumulate Heat, and pregnant women should not consume.

Ingredients

- 5 g ginseng
- 15 g Huang Qi
- 10 g Wu Wei Zi
- 8 dried longans (shelled and de-seeded)
- 1 piece beef shin (about 250 g)

Seasoning

- 1/2 tsp sea salt

Method

1 Cut beef into chunks. Blanch in boiling water. Drain and set aside. Soak and rinse ginseng, Huang Qi, Wu Wei Zi and dried longans in water.
2 Put all ingredients into a double-steaming pot. Pour in 700 ml of boiling water. Double-steam for 3 hours. Season with salt. Serve.

Chicken soup with Huai Shan, chestnuts, walnuts and dried longans

(3 to 4 servings) Ref. p.104

This soup is tasty and aromatic. It benefits the brain and strengthens the body in general. However, those with fever due to common cold or flu and those with weak Spleen and Stomach should avoid.

Ingredients

- 60 g Huai Shan (dried yam)
- 100 g chestnuts
- 80 g shelled walnuts
- 10 g dried longans (shelled and de-seeded)
- 6 red dates
- 1 free-range chicken
- 3 slices ginger

Seasoning

- 1/2 tsp sea salt

Method

1 Dress the chicken and rinse well. Chop into pieces. Blanch in boiling water. Drain. Blanch chestnuts in boiling water. Drain and peel them.
2 De-seed the red dates. Soak and rinse Huai Shan. Rinse the dried longans.
3 Put all ingredients into a pot. Add 1.7 litre of water. Bring to a vigorous boil. Turn to low heat and cook for 2 hours. Season with salt. Serve.

Black Goji and mulberry tea (2 servings)

Ref. p.107

This tea is sweet and sour with a tangy taste. It brightens the skin, darkens hair, and defies ageing. It is good for all body types, but those with diarrhoea due to Spleen-Asthenia should avoid. Pregnant women should use with care.

Ingredients

- 10 g dried black Goji berries
- 30 g dried mulberries

Seasoning

- 1 tbsp honey

Method

1 Put all ingredients into a teapot. Pour in boiling water and swirl to rinse them. Drain.
2 Pour in 500 ml of boiling water. Cover the lid and leave it for 10 minutes. Stir in honey. Strain and serve.

Honeysuckle and Xia Ku Cao tea

(2 to 3 servings) Ref. p.110

This is a popular herbal tea and is good for both sexes. However, those with Qi-Asthenia, overwhelming Dampness, Stomach- or Spleen-Asthenia may suffer from diarrhoea and may see their symptoms worsened if they consume this tea too often. Women should avoid during menstruation and pregnancy.

Ingredients

- 30 g Xia Ku Cao
- 10 g honeysuckle

Seasoning

- 30 g rock sugar

Method

1 Rinse Xia Ku Cao and honeysuckle.
2 Boil Xia Ku Cao in 1 litre of water for 45 minutes. Add honeysuckle and rock sugar. Boil for 5 minutes. Serve.

Chrysanthemum tea with Jue Ming Zi

(2 servings) Ref. Ref. p.112

This tea is sweet in taste with a hint of bitterness. Those suffering from hepatitis and high blood pressure will benefit from it. However, those with Asthenia-Coldness in the Spleen and Stomach, diarrhoea due to Spleen-Asthenia, low blood pressure and pregnant women should not consume.

Ingredients

- 20 g fried Jue Ming Zi
- 6 g chrysanthemum

Method

1 Put all ingredients into a teapot. Pour in boiling water and swirl to rinse well. Drain.
2 Pour in 500 ml of boiling water. Cover the lid. Leave it for 10 minutes. Strain and serve.

Dried pearl clam soup with Huai Shan and Goji berries (3 to 4 servings) Ref. p.114

This soup is sweet and tasty, a great tonic without building Dryness in the body. It is great for those with eye fatigue. However, those having fever due to common cold or flu and gout patients should not consume.

Ingredients

- 50 g Huai Shan
- 6 g Goji berries
- 1 carrot
- 2 slices ginger
- 80 g dried pearl clams
- 1 chicken breast

Seasoning

- 1/2 tsp sea salt

Method

1 Soak and rinse Huai Shan, Goji berries and dried pearl clams in water. Peel and cut carrot into chunks. Slice the chicken breast. Blanch chicken and dried pearl clams in boiling water together.
2 Put all ingredients into a double-steaming pot. Pour in 700 ml of boiling water. Double-steam for 3 hours. Season with salt. Serve.

Chicken liver soup with Ye Xiang Hua and Goji berries (3 to 4 servings) Ref. p.116

This soup is fragrant and delicious. Apart from improving eyesight and easing surfers' eye, it also help alleviate soreness and weakness in the tendons and bones. It is generally suitable for people with all kinds of body type. However, gout patients should replace chicken liver with lean pork.

Ingredients

- 60 g Ye Xiang Hua (cowslip flowers)
- 10 g Goji berries
- 3 to 4 sets chicken liver
- 1 tbsp finely shredded ginger

Seasoning

- 1/2 tsp sea salt

Method

1 Rinse Ye Xiang Hua and Goji berries. Rinse the chicken liver and cut into pieces.
2 Boil chicken liver and shredded ginger in 800 ml of water for 20 minutes. Add Goji berries and Ye Xiang Hua. Boil for 5 more minutes. Season with salt. Serve.

Nerve-calming tea with lily bulbs, wheat groats and lotus seed cores (2 to 3 servings) Ref. p.119

This tea is sweet and delicious. It is good for all ages. Only those with coughs due to Wind-Coldness, haemorrhagic diseases due to Asthenia-Coldness, or fever due to common cold or flu should avoid.

Ingredients

- 20 g dried lily bulbs
- 30 g wheat groats
- 5 g lotus seed cores
- 6 red dates

Seasoning

- 50 g rock sugar

Method

1 Soak and rinse the lily bulbs and wheat groats. Rinse the lotus seed cores. De-seed the red dates.
2 Put all ingredients into a pot. Add 1 litre of water and bring to a vigorous boil. Turn to medium-low heat and cook for 1 hour. Add rock sugar. Cook until it dissolves. Serve.

Chamomile tea (2 servings) Ref. p.122

This tea is fragrant and tasty. It nourishes the skin, alleviates PMS and improve sleep quality. However, pregnant women should avoid.

Ingredients

- 10 g chamomile

Method

1 Put chamomile into a teapot. Pour in boiling water. Swirl to rinse well. Drain.
2 Pour 500 ml of boiling water into a teapot. Cover the lid and leave it for 5 minutes. Strain and serve.

Date seed and arborvitae seed tea

(2 servings) Ref. p.124

This tea is sweet and aromatic with a hint of tartness. It is a great remedy for those suffering from insomnia or Yin- and Blood-Asthenia, especially novice mom who find it hard to fall asleep after delivery. However, pregnant women and those of Phlegm-Dampness body type should not consume.

Ingredients

- 15 g fried date seeds
- 15 g arborvitae seeds

Seasoning

- 1 tbsp honey

Method

1 Put both date seeds and arborvitae seeds into a teapot. Add boiling water and swirl to rinse them. Drain.
2 Pour in 500 ml of boiling water. Cover the lid and leave it for 10 minutes. Stir in honey. Serve.

Lingzhi tea with dried longans

(2 servings) Ref. p.126

This tea is sweet in taste with a hint of bitterness. It helps boost brain power and calms the nerves. However, patients should not consume much after surgical operations, including Caesarean section as it may lead to bleeding. Those having fever due to common cold or flu should also avoid.

Ingredients

- 20 g black lingzhi (sliced)
- 10 dried longans (shelled and de-seeded)

Seasoning

- 1 tbsp honey

Method

1 Put all ingredients into a teapot. Pour in boiling water. Swirl to rinse them. Drain.
2 Pour in 500 ml of boiling water. Cover the lid. Leave it for 15 minutes. Serve.

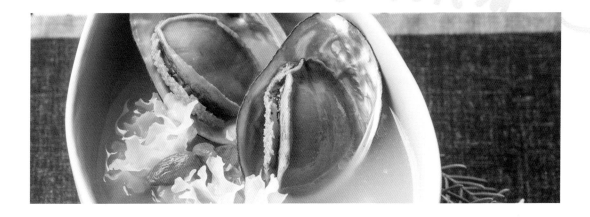

Ginseng tea with Mai Dong and Fu Shen (2 servings) Ref. p.128

This tea has strong ginseng fragrance. Those who are overworked and those suffering from chronic insomnia will benefit from it. However, common cold patients with fever, pregnant women and young children should avoid. Besides, radish is said to counteract the medicinal potency of ginseng. When you drink this tea, it's advisable to avoid consuming strong tea and radish on the same day.

Ingredients

- 5 g white ginseng (sliced)
- 10 g Mai Dong
- 15 g Fu Shen
- 8 dried longans (shelled and de-seeded)

Seasoning

- 30 g rock sugar

Method

1 Soak and rinse Mai Dong and Fu Shen in water. Rinse ginseng and dried longans.
2 Put Mai Dong, Fu Shen and dried longans into a pot. Add 800 ml of water. Boil for 1 hour. Add rock sugar and cook until it dissolves. Serve.

Abalone soup with lily bulbs and Goji berries (3 to 4 servings) Ref. p.130

This soup has great seafood flavours while nourishing the Yin and moistening Dryness. Those with Yin-Asthenia in the Liver and Kidneys, and those who have attention problem benefit from it most. However, those with fever due to common cold or flu and gout patients should not consume.

Ingredients

- 50 g dried lily bulbs
- 5 g white fungus
- 6 g Goji berries
- 2 slices ginger
- 300 g live abalones

Seasoning

- 1/2 tsp sea salt

Method

1 Soak and rinse lily bulbs and Goji berries in water. Soak white fungus in water until soft. Cut off the tough root. Scrub the abalone "feet" and shells well. Rinse and remove their innards. Blanch them in boiling water with the shells on. Set aside.
2 Put all ingredients into a pot. Add 1.5 litre of water. Bring to a vigorous boil. Turn to low heat and cook for 2 hours. Season with salt. Serve.

Lean pork soup with Ye Jiao Teng and dried scallops (2 to 3 servings) Ref. p.132

This soup does not have strong herbal taste. Those suffering from Asthenia-Fire who have to stay up till late at night. However, those with bad temper and Sthenic-Fire should avoid.

Ingredients

- 30 g Ye Jiao Teng
- 3 to 4 dried scallops
- 6 red dates
- 250 g lean pork

Seasoning

- 1/2 tsp sea salt

Method

1 Soak and rinse the Ye Jiao Teng. De-seed the red dates. Soak dried scallops in water until soft. Slice the pork and blanch in boiling water. Drain.
2 Put all ingredients into a pot. Add 1 litre of water. Bring to a vigorous boil. Turn to low heat and simmer for 1 hour 30 minutes. Season with salt. Serve.

Double-steamed pork heart soup with Fu Ling and Yuan Zhi (2 to 3 servings) Ref. p.134

This soup has a mild herbal taste and insomnia patients of any body type may consume. However, those with gastritis and gastric ulcer should use with care. Pregnant women should avoid.

Ingredients

- 15 g Fu Ling
- 6 g Yuan Zhi
- 10 g sour date kernels
- 10 dried longans (shelled and de-seeded)
- 1 piece dried tangerine peel
- 1 pork heart

Seasoning

- 1/2 tsp sea salt

Method

1 Soak and rinse Fu Ling, Yuan Zhi, sour date kernels and dried tangerine peel in water. Rinse the dried longans. Slice the pork heart thickly. Blanch in boiling water. Drain.
2 Put all ingredients into a double-steaming pot. Pour in 600 ml of boiling water. Double-steam for 3 hours. Season with salt. Serve.

Almond milk (2 servings) Ref. p.137

This drink is aromatic and nourishing. It is suitable for people with all body types. However, those with lactose intolerance and those suffering from diarrhoea should not consume.

Ingredients

- 50 g ground almond
- 500 ml milk

Seasoning

- 20 g rock sugar

Method

1 Mix ground almond with a little milk. Stir into a consistent paste. Add milk and cook over medium-low heat for 7 minutes.
2 Add rock sugar. Cook until sugar dissolves. Serve.

Hemp tea with honey (2 servings) Ref. p.140

This drink is sweet and nutty. Those who stay up at night without enough sleep and those with Asthenia-Fire uprising benefit most from it. However, pregnant women, and those with weak Spleen and Stomach or Yang-Asthenia should not consume.

Ingredients

- 20 g hemp seeds

Seasoning

- 2 tbsp honey

Method

1 Fry the hemp seeds in a dry wok until fragrant. Crush finely. Put them into a filter bag for loose tea leaves. Tie well.
2 Put the filter bag into a teapot. Add 500 ml of boiling water. Cover the lid and leave it for 10 minutes. Stir in honey and serve.

Luo Han Guo fig tea (2 servings) Ref. p.142

This sweet soup is tasty and is good for all ages. However, those suffering from diarrhoea due to Spleen-Asthenia and those with Phlegm-Dampness should consume with care.

Ingredients

- 1/2 Luo Han Guo
- 4 dried figs

Method

1 Rinse the Luo Han Guo. Crush it and set aside. Finely chop the dried figs. Fry them in a dry pan until lightly browned.
2 Put Luo Han Guo and dried figs into a teapot. Pour in 500 ml of water. Cover the lid and leave it for 15 minutes. Serve.

Creamy pumpkin sweet soup with pine nuts (3 to 4 servings) Ref. p.144

This sweet soup is tasty and is good for all ages. However, those suffering from diarrhoea due to Spleen-Asthenia and those with Phlegm-Dampness should consume with care.

Ingredients

- 200 g pumpkin
- 30 g pine nuts
- 500 ml milk

Seasoning

- 50 g red sugar

Thickening glaze (mixed well)

- 1 tbsp caltrop starch
- 1 tbsp water

Method

1 Rinse the pumpkin. Peel and de-seed. Steam in a steamer until done. Mash while still hot. Set aside. Toast the pine nuts in a dry pan until lightly browned.
2 Boil the milk. Put in mashed pumpkin and sugar. Stir well. Stir in the thickening glaze and cook until it thickens. Save in serving bowls. Sprinkled with toasted pine nuts. Serve.

Pork shin soup with night blooming cereus, almonds and carrot

(3 to 4 servings) Ref. p.146

This soup lubricates the intestines and soothes Dryness. Those suffering from constipation due to accumulated Heat in the body would benefit most. However, those of Coldness body type and those having constipation due to Qi-Asthenia should consume in moderation.

Ingredients

- 50 g night blooming cereus
- 30 g sweet almonds
- 1 carrot
- 4 dried figs
- 250 g pork shin

Seasoning

- 1/2 tsp sea salt

Method

1. Soak and rinse the night blooming cereus and almonds in water. Drain and set aside. Peel and cut carrot into chunks. Cut dried figs into halves. Cut pork shin into pieces. Blanch in boiling water. Drain.
2. Put pork shin, almonds, carrot and dried figs into a soup pot. Add 1.7 litre of water. Bring to a vigorous boil. Put in night blooming cereus. Turn to low heat and boil for 2 hours. Season with salt. Serve.

Tri-floral tea for blood circulation

(2 servings) Ref. p.149

This tea is fragrant and soothing. It helps eases tension, clears meridian congestions, improves blood flow, nourishes the skin, and lightens spots and freckles. However, pregnant women and those with loose stools should not consume.

Ingredients

- 5 g dried rose
- 5 g dried jasmine
- 5 g dried peach blossoms

Seasoning

- 1 tbsp honey

Method

1. Put all flowers into a teapot. Pour in boiling water and swirl to rinse well. Drain.
2. Pour in 500 ml of boiling water. Cover the lid and leave it for 5 minutes. Stir in honey and serve.

American ginseng tea with lotus seeds (2 servings) Ref. p.152

This tea nourishes the body and beautifies the skin. People of any body type may consume. However, those having fever due to common cold or flu and those suffering from constipation should not consume.

Ingredients

- 10 g American ginseng
- 50 g lotus seeds

Method

1. Rinse the lotus seeds. Soak in water for 1 hour. Boil them in 700 ml of water for 30 minutes.
2. Add American ginseng. Turn off the heat. Cover the lid and leave it for 10 minutes. Strain and serve.

Black rice tea with dried longans and red dates (2 servings) Ref. p.154

This tea is nourishing and beautifies the skin. It is delicious while preventing nervous prostration. It is good for all ages. However, those with Phlegm-Dampness, common cold or flu should avoid.

Ingredients

- 20 g dried longans (shelled and de-seeded)
- 8 red dates
- 30 g black rice

Method

1 Fry the black rice in a dry wok until fragrant. De-seed red dates. Slice them.
2 Put all ingredients into a teapot. Pour in 500 ml of boiling water. Cover the lid and leave it for 15 minutes. Strain and serve. You may keep on adding water until the tea goes bland.

Beef soup with beetroot, tomatoes and kidney beans (3 to 4 servings) Ref. p.156

This soup is sweet and delicious. Regular consumption helps add a healthy blush to the cheeks. It is good for all body types. But those with stomach ulcers, stomach gas and excessive stomach acid should use with care.

Ingredients

- 150 g beetroot
- 150 g tomatoes
- 50 g red kidney beans
- 250 g beef

Seasoning

- 1/2 tsp sea salt

Method

1 Peel and cut beetroot into pieces. Peel tomatoes and slice them. Soak red kidney beans in water for 1 hour. Slice the beef and blanch in boiling water. Drain.
2 Boil beef and red kidney beans in 1.5 litre of water for 1 hour. Add beetroot and tomatoes. Boil for 10 more minutes. Season with salt and serve.

Pork shin soup with cordyceps flowers, white fungus and Goji berries

(3 to 4 servings) Ref. p.158

This soup is tasty and fragrant. It's neutral in nature without being Hot or Cold. Those who are over- or underweight and those with impaired immune resistance may consume. However, those with fever due to common cold or flu should consume in moderation.

Ingredients

- 30 g cordyceps flowers
- 15 g white fungus
- 10 g Goji berries
- 6 red dates
- 250 g pork shin

Seasoning

- 1/2 tsp sea salt

Method

1 Soak and rinse cordyceps flowers, white fungus and Goji berries in water. Cut off the hard root of the white fungus. De-seed the red dates. Cut pork into chunks. Blanch in boiling water. Drain.
2 Put white fungus, red dates and pork shin into a pot. Add 1.5 litre of water. Boil for 1 hour 30 minutes. Add cordyceps flowers and Goji berries. Boil for 10 minutes. Season with salt. Serve.

Fishwort tea with bellflower root

(1 to 2 servings) Ref. p.161

This tea helps strengthen the immune system, suppresses inflammatory response, kills bacteria and fights viruses. It also improves elasticity of capillary blood vessels, promotes the regeneration of tissue, and protects the respiratory tract. However, those with Asthenia-Coldness body type should avoid.

Ingredients

- 50 g fresh fishwort
- 6 g dried bellflower root

Method

1 Rinse the fishwort. Cut into short lengths. Rinse the dried bellflower root.
2 Boil 500 ml of water. Put in fishwort and bellflower root. Cook for 20 minutes. Strain and serve.

Kudzu flower tea (1 serving) Ref. p.164

This tea helps disintoxicate alcohol. Frequent drinkers, those consuming excessive amount of alcohol and those born with Phlegm-Dampness body type would benefit most from it. However, those with Asthenia-Coldness in the Spleen and Stomach should use with care.

Ingredients

- 5 g dried kudzu flowers

Method

1 Put kudzu flowers into a teapot. Pour in boiling water and swirl to rinse well. Drain.
2 Pour in 250 ml of boiling water. Cover the lid and leave it for 5 minutes. Serve.

Egg drop soup with pork and common purslane (3 to 4 servings) Ref. p.166

This soup is fragrant with a hint of sourness. Regular consumption helps prevent pulmonary sarcoidosis and pneumoconiosis. However, those with Asthenia-Coldness in the Spleen and Stomach, diarrhoea and pregnant women should avoid.

Ingredients

- 250 g fresh common purslane
- 2 eggs
- 300 g lean pork

Seasoning

- 1/2 tsp sea salt

Method

1 Rinse the purslane and cut into short lengths. Crack and whisk the eggs. Rinse the pork and slice it. Blanch it in boiling water. Drain.
2 Cook lean pork in 800 ml of boiling water. Bring to a vigorous boil again. Put in the purslane. Boil for 20 minutes. Stir in whisked eggs at last. Season with salt. Cook until the strands of egg ribbons float to the top. Serve.

Snakehead fish soup with yellowcress and candied dates (3 to 4 servings) Ref. p.168

This soup is nourishing, yet without feeling greasy. It is neutral in nature without being Hot or Cold. It is good for all ages and all body types. Only those with Asthenia-Coldness in the Spleen and Stomach should consume in moderation.

Ingredients

- 300 g fresh variable leaf yellowcress
- 2 candied dates
- 2 slices ginger
- 1 to 2 snakehead fish

Seasoning

- 1/2 tsp sea salt

Method

1 Cut off the roots of the yellowcress. Rinse and cut into short lengths. Rinse the candied dates.
2 Dress the snakehead fish. Rinse well and wipe dry. Fry in a little oil on both sides until golden.
3 Boil 1.7 litre of water. Put in ginger, snakehead fish and candied dates. Bring to a vigorous boil. Put in yellowcress. Turn to low heat and simmer for 2 hours. Season with salt. Serve.

Dried oyster soup with tofu skin, water chestnuts and hairy moss

(3 to 4 servings) Ref. p.170

This soup is nourishing and tasty. It also regulates intestinal functions and cleanses the respiratory tract. Those who work in dusty environment with poor air quality may consume regularly. Generally, it is good for all body types. Only those with Asthenia-Coldness in the Spleen and Stomach and those with skin problems should consume in moderation.

Ingredients

- 1 sheet dried tofu skin
- 10 water chestnuts
- 10 g black hairy moss
- 100 g dried oysters
- 3 slices ginger
- 250 g lean pork

Seasoning

- 1/2 tsp sea salt

Method

1 Rinse the tofu skin. Peel and rinse the water chestnuts. Cut into halves. Soak and rinse the black hairy moss in water.
2 Rinse the soak the dried oysters in water. Slice the pork. Blanch dried oysters and pork in boiling water. Drain.
3 Put all ingredients into a pot. Add 1.7 litres of water. Bring to a vigorous boil. Turn to low heat and simmer for 2 hours. Season with salt. Serve.

Lemongrass and apple tea

(2 servings) Ref. p.173

This tea is aromatic and fruity in taste. It warms the stomach and expels Coldness. Regular consumptions help prevent high cholesterol level and gallstones. However, those with Yin-Asthenia and accumulated Heat should not consume much.

Ingredients

- 3 stems lemongrass
- 1 apple
- 4 slices ginger

Seasoning

- 1 tbsp red sugar

Method

1 Rinse and slice the lemongrass. Peel and core the apple. Dice it.
2 Boil 700 ml of water in a pot. Put in lemongrass, ginger and diced apple. Boil for 20 minutes. Add sugar and cook until sugar dissolves. Serve.

Honeysuckle and Dang Gui Tea

(2 servings) Ref. p.176

This tea is sweet in taste with a hint of bitterness. It helps prevent vein thrombosis. However, those with Asthenia-Coldness in the Spleen and Stomach, those having loose stool and diarrhoea, and pregnant women should avoid.

Ingredients

- 10 g honeysuckle
- 10 g Xuan Shen
- 6 g Dang Gui
- 3 g liquorice

Method

1 Soak and rinse all ingredients.
2 Put in a pot and add 700 ml of water. Boil for 20 minutes. Strain and serve.

Red date ginger tea with cassia branches (2 servings) Ref. p.178

This tea expels Coldness, boosts Qi (vital energy) flow in the body, and alleviates numbness and pain due to rheumatism. However, those with Yin-Asthenia accompanied by accumulated Heat, those with purple patches on the skin due to overwhelming Heat in the Blood, and pregnant women should avoid.

Ingredients

- 6 g cassia branches
- 6 g dried ginger
- 6 red dates

Seasoning

- 1 tbsp honey

Method

1 De-seed red dates and slice them. Rinse all ingredients.
2 Put all ingredients into a pot. Add 600 ml of water. Boil for 15 minutes. Let cool slightly. Stir in honey. Serve.

Kidney-boosting soup with Du Zhong and walnuts (3 to 4 servings) Ref. p.180

This soup benefits the Liver and Kidneys, while strengthening the bones and sinews. Those who have to stand in air-conditioned environment for prolonged hours may consume regularly. Though it is suitable for all people in general, those suffering from gout or high blood cholesterol level should replace the pork kidneys with 250 g of lean pork.

Ingredients

- 30 g Du Zhong
- 6 g Goji berries
- 50 g shelled walnuts
- 20 g cooked Di Huang
- 3 slices ginger
- 1 pair pork kidneys

Seasoning

- 1/2 tsp sea salt

Method

1 Rinse Du Zhong, Goji berries, walnuts and cooked Di Huang separately. Tear the white membrane off the kidneys as much as you can. Slice the kidneys and rinse in running water. Blanch in boiling water. Drain.
2 Put Du Zhong, walnuts, cooked Di Huang, ginger and pork kidneys in a soup pot. Add 1.5 litre of water. Boil for 1 hour. Add Goji berries and boil for 5 more minutes. Season with salt. Serve.

Chicken soup with Ji Xue Teng and red dates (3 to 4 servings) Ref. p.182

This soup activates meridian flow, strengthens the sinews, promotes blood cell regeneration and warms the lower back and knees. Those who work in air-conditioned environment for prolonged hours would benefit from it most. However, those with Yin-Asthenia and overwhelming Fire should use with care.

Ingredients

- 30 g Ji Xue Teng
- 8 red dates
- 3 slices ginger
- 250 g chicken breast meat

Seasoning

- 1/2 tsp sea salt

Method

1 Rinse and soak Ji Xue Teng in water. De-seed red dates. Rinse the chicken and cut into chunks. Blanch in boiling water. Drain.
2 Put all ingredients into a pot. Add 1.5 litre of water. Bring to a vigorous boil. Turn to low heat and cook for 1 hour. Season with salt. Serve.

Wood ear fungus tea with ginger and red dates (2 servings) Ref. p.185

This tea is rich in collagen, iron and calcium etc. It is a great drink for women and others to prevent anaemia and osteoporosis. However, those with haemorrhagic diseases or diarrhoea should not consume. Pregnant women should consume in moderation.

Ingredients

- 20 g dried black wood ear fungus
- 30 g ginger
- 6 red dates

Seasoning

- 30 g red sugar

Method

1 Soak wood ear fungus in water until soft. Cut off the root and then cut into small pieces. Shred the ginger. De-seed the red dates.
2 Put all ingredients into a pot. Add 700 ml of water and boil for 30 minutes. Add sugar and cook until it dissolves. Serve.

Coralline algae tea with Goji berries (2 servings) Ref. p.188

Regular consumption of this tea helps promote blood circulation, eliminates faeces trapped in the folds of the intestines, and detoxifies. However, coralline algae is Cold in nature, so that those with Yang-Asthenia body type should consume in moderation.

Ingredients

- 15 g coralline algae
- 5 g Goji berries

Seasoning

- 50 g rock sugar

Method

1 Soak coralline algae in water until completely transparent. Rinse and cut into short lengths. Soak and rinse Goji berries.
2 Cook coralline algae in 600 ml of water for 20 minutes. Add Goji berries and cook for 5 more minutes. Add rock sugar and cook until it dissolves. Serve.

Jiao Gu Lan tea with honey

(2 servings) Ref. p.190

This tea is fragrant with a grassy freshness. It is effective in maintaining the immune resistance and strengthening positive energy, while preventing and curing cancer at the same time. However, in rare cases, it may lead to nausea, vomiting or dizziness. Those with Asthenia-Coldness in the Spleen and Stomach should consume with care.

Ingredients

- 10 g Jiao Gu Lan

Seasoning

- 1 tbsp honey

Method

1. Put Jiao Gu Lan into a teapot. Pour in boiling water and swirl to rinse well. Drain.
2. Add 500 ml of boiling water. Cover the lid and leave it for 10 minutes. Stir in honey. Serve.

Tortoise shell soup with Tu Fu Ling, small red beans and Job's tears (3 to 4 servings) Ref. p.192

This is a popular healthful soup that clears Heat and detoxifies. Those who work in environment with poor air quality should consume regularly. However, those with Yang-Asthenia, or Asthenia-Coldness in the Spleen and Stomach, and pregnant women should avoid.

Ingredients

- 60 g fresh Tu Fu Ling
- 30 g small red beans
- 30 g Job's tears
- 2 candied dates
- 200 g dried tortoise shell
- 250 g lean pork

Seasoning

- 1/2 tsp sea salt

Method

1. Slice the Tu Fu Ling. Soak in water briefly and rinse well. Soak and rinse the small red beans and Job's tears in water.
2. Soak and rinse the tortoise shell. Blanch in boiling water. Drain. Slice the pork. Rinse and blanch in boiling water. Drain.
3. Put all ingredients into a pot. Add 1.7 litres of water. Bring to a vigorous boil. Turn to low heat and cook for 2 hours. Season with salt. Serve.

Chicken soup with monkey head mushrooms and shiitake mushrooms

(3 to 4 servings) Ref. p.194

This soup detoxifies and beautifies the skin. It also strengthen the body, promotes metabolism and helps excrete metabolic waste. However, those with Coldness-Dampnes in the Spleen and Stomach, poor Qi (vital energy) flow, and those with itchy skin should avoid.

Ingredients

- 3 dried monkey head mushrooms
- 5 – 6 dried shiitake mushrooms
- 5 g algae (such as spirulina)
- 3 slices ginger
- 1 free-range chicken

Seasoning

- 1/2 tsp sea salt

Method

1. Soak and rinse monkey head mushrooms, shiitake mushrooms and algae in water separately. Cut off the stems from monkey head mushrooms and shiitake mushrooms.
2. Dress and rinse the chicken. Chop into large chunks. Blanch chicken and monkey head mushrooms in boiling water. Drain.
3. Put all ingredients into a soup pot. Add 1.7 litre of water. Bring to a vigorous boil. Turn to low heat and cook for 2 hours. Season with salt. Serve.

飲出好體質
上班族每天養生法

Drinks and Soups for Good Health
Daily Tonics for Busy Working Classes

作者	Author
芳姐	Cheung Pui Fong
策劃/編輯	Project Editor
譚麗琴	Catherine Tam
攝影	Photographer
細權	Leung Sai Kuen
美術設計	Art Design
鍾啟善	Nora Chung

出版者 Publisher

Forms Kitchen

香港鰂魚涌英皇道1065號 Room 1305, Eastern Centre, 1065 King's Road,
東達中心1305室 Quarry Bay, Hong Kong.
電話 Tel: 2564 7511
傳真 Fax: 2565 5539
電郵 Email: info@wanlibk.com
網址 Web Site: http://www.wanlibk.com
　　　　http://www.facebook.com/wanlibk

發行者 Distributor

香港聯合書刊物流有限公司 SUP Publishing Logistics (HK) Ltd
香港新界大埔汀麗路36號 3/F., C&C Building, 36 Ting Lai Road,
中華商務印刷大廈3字樓 Tai Po, N.T., Hong Kong
電話 Tel: 2150 2100
傳真 Fax: 2407 3062
電郵 Email: info@suplogistics.com.hk

承印者 Printer

Best Motion

出版日期 Publishing Date
二零一七年一月第一次印刷 First print in January 2017
二零一八年二月第二次印刷 Second print in February 2018